$40.95
Y0-AAF-639

MATH/STAT
LIBRARY

COMPLEX SYSTEMS DYNAMICS

AN INTRODUCTION TO AUTOMATA NETWORKS

Gérard Weisbuch
Ecole Normale Supérieure

Translated by
Sylvie Ryckebusch
California Institute of Technology

Lecture Notes Volume II

SANTA FE INSTITUTE
STUDIES IN THE SCIENCES OF COMPLEXITY

Addison-Wesley Publishing Company
The Advanced Book Program
Redwood City, California • Menlo Park, California • Reading, Massachusetts
New York • Don Mills, Ontario • Wokingham, United Kingdom • Amsterdam
Bonn • Sydney • Singapore • Tokyo • Madrid • San Juan

Publisher: *Allan M. Wylde*
Production Manager: *Jan V. Benes*
Marketing Manager: *Laura Likely*

Director of Publications, Santa Fe Institute: *Ronda K. Butler-Villa*
Technical Assistant, Santa Fe Institute: *Della L. Ulibarri*

Originally published in 1989 as *Dynamique Des Systèmes Complexes:* une Introduction aux Reséaux d'automates, by InterEditions, France.

Library of Congress Cataloging-in-Publication Data

Weisbuch, G.
(Dynamique des systèmes complexes. English)
Complex systems dynamics : an introduction to automata networks / Gérard Weisbuch : translated by Sylvie Ryckebusch.
p. cm.—(Santa Fe Institute studies in the sciences of complexity. Lecture notes : v. 2)
Translation of: Dynamique des systèmes complexes.
Includes bibliographical references (p.) and index.
1. Computational complexity. 2. Machine theory. I. Title. II. Series.
QA267.7.W4513 1990 511.3-dc20 90-1221
ISBN 0-201-52887-8

This volume was typeset using TEXtures on a Macintosh II computer. Camera-ready output from an Apple LaserWriter Plus Printer.

Copyright © 1991 by Addison-Wesley Publishing Company, The Advanced Book Program, 350 Bridge Parkway, Redwood City, CA 94065

All rights reserved. No part of this publication may be reproduced, stored in a retrieval system, or transmitted in any form or by any means, electronic, mechanical, photocopying, recording, or otherwise, without the prior written permission of the publisher. Printed in the United States of America. Published simultaneously in Canada.

ABCDEFGHIJ-MA-943210

About the Santa Fe Institute

The *Santa Fe Institute* (SFI) is a multidisciplinary graduate research and teaching institution formed to nurture research on complex systems and their simpler elements. A private, independent institution, SFI was founded in 1984. Its primary concern is to focus the tools of traditional scientific disciplines and emerging new computer resources on the problems and opportunities that are involved in the multidisciplinary study of complex systems—those fundamental processes that shape almost every aspect of human life. Understanding complex systems is critical to realizing the full potential of science, and may be expected to yield enormous intellectual and practical benefits.

All titles from the *Santa Fe Institute Studies in the Sciences of Complexity* series will carry this imprint which is based on a Mimbres pottery design (circa A.D. 950–1150), drawn by Betsy Jones.

Santa Fe Institute Editorial Board
August 1990

Dr. L. M. Simmons, Jr., *Chair*
Executive Vice President, Santa Fe Institute

Professor Philip W. Anderson
Department of Physics, Princeton University

Professor Kenneth J. Arrow
Department of Economics, Stanford University

Professor W. Brian Arthur
Dean & Virginia Morrison Professor of Population Studies and Economics, Food Research Institute, Stanford University

Dr. David K. Campbell
Director, Center for Nonlinear Studies, Los Alamos National Laboratory

Dr. George A. Cowan
President, Santa Fe Institute and Senior Fellow Emeritus, Los Alamos National Laboratory

Professor Marcus W. Feldman
Director, Institute for Population & Resource Studies, Stanford University

Professor Murray Gell-Mann
Division of Physics & Astronomy, California Institute of Technology

Professor John H. Holland
Division of Computer Science & Engineering, University of Michigan

Dr. Bela Julesz
Head, Visual Perception Research, AT& T Bell Laboratories

Professor Stuart Kauffman
School of Medicine, University of Pennsylvania

Dr. Alan S. Perelson
Theoretical Division, Los Alamos National Laboratory

Professor David Pines
Department of Physics, University of Illinois

Santa Fe Institute Studies in the Sciences of Complexity

PROCEEDINGS VOLUMES

Volume	Editor	Title
I	David Pines	Emerging Syntheses in Science, 1987
II	Alan S. Perelson	Theoretical Immunology, Part One, 1988
III	Alan S. Perelson	Theoretical Immunology, Part Two, 1988
IV	Gary D. Doolen et al.	Lattice Gas Methods of Partial Differential Equations, 1989
V	Philip W. Anderson et al.	The Economy as an Evolving Complex System, 1988
VI	Christopher G. Langton	Artificial Life: Proceedings of an Interdisciplinary Workshop on the Synthesis and Simulation of Living Systems, 1988
VII	George I. Bell & Thomas G. Marr	Computers and DNA, 1989
VIII	Wojciech H. Zurek	Complexity, Entropy, and the Physics of Information, 1990
IX	Alan S. Perelson & Stuart A. Kauffman	Molecular Evolution on Rugged Landscapes: Proteins, RNA and the Immune System, 1990

LECTURES VOLUMES

Volume	Editor	Title
I	Daniel L. Stein	Lectures in the Sciences of Complexity, 1988
II	Erica Jen	1989 Lectures in Complex Systems

LECTURE NOTES VOLUMES

Volume	Editor	Title
I	John Hertz et al.	Introduction to the Theory of Neural Computation, 1990
II	Gérard Weisbuch	Complex Systems Dynamics

Contents

L.M. Simmons, Jr.

Foreword

We are witnessing the creation of new sciences of complexity, sciences that may well occupy the center of intellectual life in the twenty-first century. The Santa Fe Institute was founded to assist at the birth of these new sciences. Those involved in this activity are proceeding under the conviction that there is a common set of principles shared by the disparate complex systems under study, that the time is ripe to understand those principles, and that it is essential to develop them and the associated tools for dealing in a systematic way with complex systems. This lecture notes volume by Gérard Weisbuch, a frequent visitor to the Institute, introduces the reader to a rich array of tools and concepts that are central to understanding and modeling complex systems.

Complex systems typically do not fit within the confines of one of the traditional disciplines, but for their successful study require knowledge and techniques from several disciplines. Thus one task of the Institute has been to find new ways to encourage cooperative research among scholars from different fields. The Studies in the Sciences of Complexity series is one means that the Institute has adopted for accelerating the development of the sciences of complexity. These volumes make available to the scholarly community the results of conferences and workshops sponsored by the Institute, lectures presented in the complex systems summer school, and, as in this volume, other lecture notes by active researchers.

The sciences of complexity are emerging in part as a synthesis of some of the traditional sciences, including biology, physics, and mathematics. In part they are emerging as a result of new ideas, new questions, and new techniques only recently developed. Among these latter are the emergence of heretofore undreamed of computer power on the scientist's desktop and the not unrelated progress in nonlinear dynamics, computer graphics, and adaptive programs.

This volume is a particularly appropriate addition to the series, providing an accessible introduction to complex systems viewed as networks of automata. It is intended for a broad audience ranging from physicists and computer scientists to biologists and introduces a set of techniques of equally broad applicability.

The Santa Fe Institute, and hence this series, would not exist without the support of farsighted individuals in government funding agencies and private foundations who have recognized the promise of the new approaches to complex systems research being fostered here. It is a pleasure to acknowledge the broad research grants received by the Institute from the U.S. Department of Energy (ER-FG05-88ET25054), the John D. and Catherine T. MacArthur Foundation, and the National Science Foundation (PHY-8714918) together with numerous smaller grants that have made the work of the Institute possible.

L. M. Simmons, Jr.
Santa Fe
July 31, 1990

Christof Koch

Foreword

What are the similarities between a universe populated by rapidly evolving galaxies, a network of highly interconnected digital processors, simple automata living in the two-dimensional plane, and neuronal networks in the brain? At some level, the dynamic behavior of these systems follows similar laws, dictated by the nature of the interaction between the individual elements. The study of these evolutionary laws is the primary aim of the field of *complexity*, a synthesis between mathematics, physics, neurobiology, molecular biology, and computer science. The present work is an introduction to this field, using primarily examples drawn from the physics of disordered systems (e.g., Ising model and spin glasses), neural networks, and the origins of life. It aims to highlight the similarities between these different systems, how the overall behavior changes over time and whether the system settles into a unique state, oscillates or shows phase transitions. It is a highly interdisciplinary book, at the intersection of a large number of fields (without requiring, however, the reader to be an expert in physics, mathematics, and biology). It therefore represents an excellent addition to Addison-Wesley's "Computation and Neural Systems" series.

The Computation and Neural Systems Series—Over the past 600 million years, biology has solved the problem of processing massive amounts of noisy and highly redundant information in a constantly changing environment by evolving networks of billions of highly interconnected nerve cells. It is the task of scientists—be they mathematicians, physicists, biologists, psychologists, or computer scientists—to understand the principles underlying information processing in these complex structures. At the same time, researchers in machine vision, pattern recognition, speech understanding, robotics, and other areas of artificial intelligence can profit from understanding features of existing nervous systems. Thus, a new field is emerging: the study of how computations can be carried out in extensive networks of heavily interconnected processing elements, whether these networks are carbon-

or silicon-based. Addison-Wesley's new "Computation and Neural Systems" series will reflect the diversity of this field with textbooks, course materials, and monographs on topics ranging from the biophysical modeling of dendrites and neurons, to computational theories of vision and motor control, to the implementation of neural networks using VLSI or optics technology, to the study of highly parallel computational architectures.

Christof Koch
Pasadena, California
August 15, 1990

Preface

In 1982, a paper by John Hopfield entitled "Neural Networks and Physical Systems with Emergent Collective Computational Abilities" appeared in the *Proceedings of the National Academy of Sciences.* The "recipe" of the paper was the following:

- Start with a network composed of very simplified models of the components of the nervous system.
- Apply some ideas from statistical mechanics, and rely on numerical computer simulation to obtain results.
- Compare emergent collective properties to qualitatively similar properties of the brain.
- From this, propose new circuits and new computer architectures.

Although this was not a new subject, as is evident from the bibliography, this paper and a handful of others of comparable quality were the starting point of an extraordinary rebirth of this field of scientific activity. Researchers, funding, meetings, papers, and new machines rapidly materialized in ever greater numbers.

The goal of this book is to introduce the reader to this research field in order that he or she may understand its methods and concepts as well as the many varied fields of application. Among these, let us just mention the most important:

- Physics of disordered systems and of growth and forms.
- The biology of the brain, but also certain key problems concerning development, immunology, and the origin of life.
- Signal processing and the design of parallel computers.

This book is a general introduction aimed at readers with a university education in science or engineering. An understanding of elementary calculus is required. We have attempted to avoid modern mathematical formalism, and consequently, although most algorithms are explained in the appendix, we leave it to the reader to look up long calculations and elegant proofs in the appropriate references. In fact the work of the researchers in this field often involves numerical simulation, which is a "hybrid" of theory and experiment.

The reader will note that the subject is not only rich in methods, but also enables a number of significant concepts to be defined or stated in a different framework. All of these new concepts appear in italics in the paragraph of the book in which they are first defined. This paragraph is referenced in the index. The applications which are discussed do not require that the reader be simultaneously a physicist, a biologist, and a specialist in signal processing—all of the necessary concepts are in principle redefined in the text. However, it sometimes happens that for a given problem, the performance of automata networks is compared to that of more traditional methods; in this case the explanations of the latter are very succinct.

The organization of the book is as follows: In chapter 1, we define complex systems, to which we will apply the method of automata networks. A very brief overview will place new perspectives resulting from advances in theoretical biology and parallel computation in historical context. Fundamental definitions relating to automata, networks, and their dynamics will be given in chapter 2, along with several examples. One-dimensional cellular automata networks discussed in chapter 3 will give other simple examples which enable the dynamics of the networks to be understood and characterized. Chapter 4 discusses two-dimensional cellular automata networks and their application to the physics of growth and to fluid mechanics.

Chapter 5, which is on formal neurons, is the basis for the four following chapters. The explanation of the Hopfield model includes the fundamental notions of a formal neuron, recognition, associative memories, and learning. Chapter 6 deals with higher performance methods, which go beyond certain limitations of the Hopfield model. Along the same lines, chapter 7 discusses the back-propagation technique, which is one of the most powerful techniques to date in applications of signal processing.

In chapter 8 we tackle the domain of probabilistic automata. The notion of temperature is restated in this context, which enables a link to be established between the problems of associative memories and of spin glasses. Chapter 9 widens our perspective. Simulated annealing allows a parallel approach to combinatorical optimization, which is a class of problems of great technological importance. An application to image processing is discussed.

Chapter 10 brings us back to more general considerations on organization, chaos, and the genericity of the dynamical properties of automata networks. Finally, chapter 11 discusses the application of these ideas to population genetics in the broadest sense.

In the conclusion, in chapter 12, we first discuss several limitations of the network method, and then we give a few bibliographical references for the reader who would like more detailed information about certain subjects in the book, or about methods and applications which were not discussed.

We think that the automata network approach could shed light on many areas, from biology to parallel computation, including physics. This by no means implies that it can enable us to understand everything, and even less that this book covers all aspects mentioned in the introduction. We believe that modern science is loathe to accept global concepts, and consequently, the purpose of this book is to clarify some of these concepts by showing by examples how broadly they can be applied. We believe that the enlightenment brought about by concepts from other fields can only improve the understanding of a scientific field. However, readers pressed for time can concentrate their efforts according to their own interests. Chapters 1, 2, 3, and 5 are probably indispensable. Chapters 4, 8, and 10 focus on physics and dynamical systems. Chapters 6, 7, and 9 deal with signal processing and combinatorial optimization. Chapter 8 is recommended for non-physicists, as it will help them to understand chapter 9. Chapter 11 deals with developmental biology and evolution.

We have not attempted to give a complete bibliography, as the number of papers published in this field is already of the order of a few thousand. We do not even claim to have listed the most important works. The references are either remarkably clear books or papers, articles which inspired a passage in the book, or reference works which allow the reader to delve into a particular field.

A few references have been added to the English version.

The choice of discussed subjects reflects our own research interests. Many researchers with whom we have collaborated deserve thanks, whether or not they are at the Statistical Physics Laboratory of the Ecole Normale Supérieure: Henri Atlan, Bernard Derrida, Françoise Fogelman-Soulié, Dominique d'Humières, Werner Krauth, Jean-Pierre Nadal, Gérard Toulouse, and Jean Vannimenus. Their suggestions and corrections have greatly contributed to this book. The constant encouragement of Michèle Leduc, director of the French series where this book was originally published, was invaluable throughout the preparation of this manuscript, from start to finish. Thanks are also due to Sylvie Ryckebusch and Ronda Butler-Villa for the English version of this book.

Gérard Weisbuch
Paris, July 1990

ONE

Introduction

The aim of this book is to introduce you to a mathematical method and a number of concepts adapted to the description of complex systems. Although complexity is now a somewhat overused expression, it has a precise meaning in this book: *a complex system is a system composed of a large number of different interacting elements.*

Statistical physics has accustomed us to mathematical descriptions of systems with a large number of components. The thermodynamic properties of ideal gases were understood as early as the end of the 19th century, while those of solids were understood at the beginning of the 20th century. In both cases, two important properties make modeling easy:

- These are systems in which all of the components are identical.
- If the interactions between the components are very weak, they can be ignored, as in the case of ideal gases. Otherwise, as in the case of solids, we can use linearization methods to put the problem into a form in which these simplifications can be made.

These systems are therefore not considered complex (see table 1.1).

On the other hand, here are some examples of complex systems, several of which will be studied in this book:

- The human brain is composed of approximately ten billion cells, or neurons. These cells interact by means of electrochemical signals through their synapses. Even though there may not be very many different types of neurons, they differ in the structure of their connections.
- Computer systems use large numbers of different electronic components, at the level of transistors or logic gates.

TABLE 1.1 Overview of approaches to multicomponent dynamical systems. We can see from this admittedly crude table that automata networks allow us to deal with the least favorable case, that of large numbers of different interacting components.

Approaches	Number of Components	Interactions	Components	Applications
Linear Systems	Large	Weak	Different	Wave Physics
Statistical Mechanics	Large	Strong or Weak	Identical	Ferromagnetism, Phase Transitions
Systems of Nonlinear Differential Equations	Small	Strong or Weak	Different	Nonlinear Oscillations
Automata Networks	Large	Strong or Weak	Different	Complex Systems

- Multiphase systems in physical chemistry have several physical or chemical phases: there are many different such systems, from mayonnaise to composite materials used in aeronautics.
- Social and economic systems, of course, are composed of different interacting elements.

In fact, the great majority of natural or artificial systems are of a complex nature, and scientists choose more often than not to work on systems simplified to a minimum number of components, which allows him or her to observe "pure" effects. In short, this is the Cartesian approach.

We will attempt in this book to describe another approach, which is to simplify as much as possible the components of a system, so as to take into account their large number. This idea has emerged from a recent trend in research known to physicists as *the physics of disordered systems.*

1–1 Disordered Systems

As we pointed out earlier, a large class of physical systems, known as multiphase systems, are disordered at the macroscopic level, but some are disordered even at the microscopic level. Glasses, for example, differ from crystals in that interatomic bonds in a glass are not distributed according to the symmetries which we observe in crystals. In spite of this disorder, the macroscopic physical properties of a glass of a given composition are generally the same for different samples, as for crystals.

In other words, disorder in a system does not lead to random behavior. The simple models used by physicists are based on periodic networks, or grids, and simplified components of two different types are placed on the nodes, such as conductors and insulators in the problem known as percolation. These components are randomly distributed, and the interactions are limited to pairs of neighboring nodes. For large enough networks, we perceive that certain interesting properties do not depend on the particular sample created by a random selection, but of the parameters of this selection. In the case of the aforementioned insulator/conductor mixture, the conductivity between the two edges of the sample depends only on the ratio of the number of conductive sites to the number of insulating sites.

Many important concepts, as well as the use of massive computer simulations, were introduced in this framework by physicists, but the motivations for the automata network approach were formulated by the first computer scientists in the forties.

1–2 Origins

Like many recent inventions, practical computer science was born from the research efforts deployed during World War II. At the time, the computer was still in its infancy, and it was not obvious that the future machine would be very different from the human brain both in terms of its architecture and function. Recall that the first human attempts at flight used as a model the flight of birds. In 1943, two scientists, the mathematician Pitts and the neurophysiologist McCulloch, proposed a system composed of *formal neurons*, that is to say simple units that had logical rules of behavior for transmission and processing of information which were inspired by neurons in the brain (as they were perceived at the time). They showed that the computational potential of this machine was equivalent to that of the theoretical "reference" machine of computer scientists, the Turing machine. The advantage of such an equivalence is that the properties of the Turing machine have been the object of numerous studies, and are clearly established.

Around the same time, the "father" of the modern computer, John von Neumann, was attempting to solve one of the paradoxes of living systems, that of self-reproduction: is it possible to construct an artificial system capable of reproducing itself? Of course, it would have been prohibitively expensive, even for an American research group, to construct a robot capable of operating the machine tools needed to machine and assemble its own parts. However, the solution of the paradox did not require a true mechanical implementation, but simply a logical proof that such a robot could be built. Von Neumann therefore reduced the real world to a model world by drawing a grid (albeit infinite), each square containing little mathematical beings, or cellular automata, which lived and died. Each square contained one automaton, representing a cell of this artificial organism. The logical solution to the problem resided in the choice of interaction rules between the automata and the search for the configurations of cells which could multiply during the "life" of the organism.

These two primeval examples show the approach taken by theorists in the field of automata networks:

- We choose to oversimplify the components of the system whose global behavior we would like to model. The formal neurons introduced by McCulloch and Pitts are a cartoon-like simplification of living neurons, the object of modern electrophysiology. Likewise, there exists no organism which has a structure even remotely resembling von Neumann's cellular automata.
- Nonetheless, these simplifications enable us to apply rigorous methods and to obtain exact results.
- This approach is a dynamical approach. As in the differential equations methods, we start from a *local description* of the system, in terms of the short-term state changes of the components as a result of their interactions. We expect the *global description* of the system from the method, that is to say the long-term behavior of the system as a whole. The global behavior can be very complex, and it can be interpreted in terms of *emergent properties.* Contained in this notion is the idea that the properties are not *a priori* predictable from the structure of the local interactions, and that they are of functional significance, whether one is talking about biological modeling or technological devices.

1–3 Toward a Technology

The automata network model enables natural complex systems to be described semi-quantitatively. It is also the basis for technological applications in signal processing, pattern recognition, artificial intelligence, etc. which we will discuss further in this book. This method circumvents two of the obstacles confronted by modern computer science.

The first such obstacle is hardware related. The recent progress in lowering costs, speeding up processing and increasing memory capacity of computers can be attributed to the increasing miniaturization of electronic devices. However, the current dimensions of these devices are approaching limits imposed by quantum mechanics—at least one electron is needed for each bit of information. What's more, the speed of information transfer in the connections is limited by the speed of light. It follows that we cannot keep extrapolating into the future the exponential increase in performance of recent years. We are approaching saturation.

The second obstacle has to do with software. Current processors have become very complex, and design time has increased accordingly. The development of software is generally far behind that of hardware, while the impressive capabilities of modern hardware are not being fully exploited.

In order to speed up calculations even more without increasing the density of integrated circuits, the solution is to use a large number of processors to do several computations in parallel. The hope is that in this manner the processing time will be divided by the number of processors. Each processor can be considered an automaton, while the set of processors constitutes a network. This is the field of *parallel computation*.

Some would object that we have jumped over a hole only to fall into a chasm; indeed, it is much more difficult to program a parallel computer than a classical sequential machine. In a classical computer, the operations are programmed one after the other. In a parallel computer, the timing of the cooperation as well as the information exchange between the processors must be managed. As we will see later, for many computationally intensive, but nonetheless parallelizable tasks, the difficulty of programming can be avoided by a learning technique. Instead of listing the series of operations to be executed by the machine, it suffices to present it with some examples, along with some general guidelines about how they are to be treated. In many cases, this approach is much simpler to implement than any other programming approach.

Now that we have discussed the motivations, in chapter 2 we will explain the essential concepts related to automata networks and their dynamics. The two following chapters, dedicated to cellular automata, will give many simple examples and are relevant to applications in fluid mechanics. We will begin in chapter 5 to answer some of the questions which were raised in this introduction.

TWO

Definitions

Intuitively, a network of automata can be thought of as a set of interacting elements. We will return to using discrete mathematics in order to simplify the elements of this set as much as we can. We therefore abandon continuous mathematics for the time being. Let's begin with time: instead of being a real variable, as it is in physics, in this model, time is broken up into even intervals numbered from 1 to n. The variables, represented by automata, are updated during each interval. Therefore it makes sense to represent time by a whole number:

$$t = \{0, 1, 2, 3, \ldots, n, \ldots\}.$$

Note that in a computer, time is also broken up into intervals by the clock, which coordinates the changes of state of the logic units.

2–1 Automata

2–1–1 Classical Definition

The second discretization operation is to replace the continuous variables and differential equations by finite state automata.

In computer science, an automaton is defined by three discrete sets:

1. $\mathbf{I}$, the set of *inputs* i,
2. $\mathbf{S}$, the set of *internal states* s, and
3. $\mathbf{O}$, the set of *outputs* o,

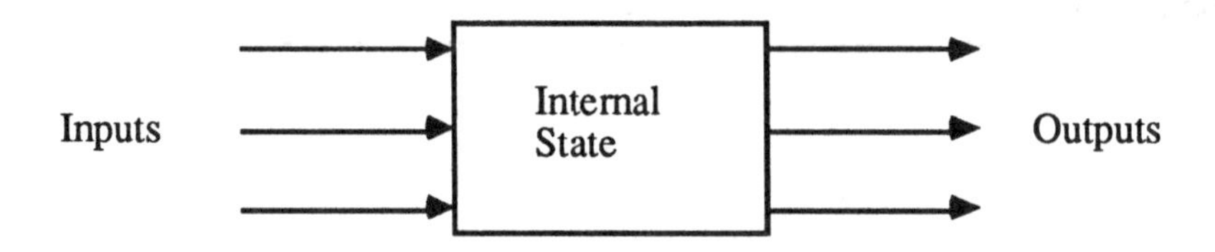

FIGURE 2.1 A computer science automaton.

and by two mappings:

1. $S(i, s)$, the ***state change function***, which maps the input and the state at time t to the new state at time $t+1$, and
2. $O(i, s)$, the ***output function***, which maps the input and the state at time t to the output at time $t+1$ (Fig. 2.1).

Of course, in the general case, there can be many inputs, internal states, and outputs, so S and O are functions of the vectors i, s, and o. The important point is that these variables must be discrete and therefore can be represented by integers or sets of integers. k, the number of inputs of an automaton, is called the *input connectivity* of that automaton.

A pulse counter, for example, can be considered an automaton (Fig. 2.2). Its two inputs are the input signal and the reset. When the reset is in state 0 at time t, the internal state at time $t+1$ is the state at time t plus the input signal at time t. The input signal can have either the value 0 or 1. If the reset is set to 1, the internal state becomes 0 at the next time step.

The output is a binary representation of the number of pulses received since the last time the internal state was zeroed (see the appendix for the relationship between binary and decimal coding). The output which corresponds to the *least significant bit* (LSB) changes to state 0 (or 1) at time $t+1$ if the internal state at time t is odd (or even). (In binary notation, a bit is more significant if it corresponds to a higher power of 2. For example, 5 is represented by 0101: the *most significant bit* (MSB) is a 0, which corresponds to the third power of 2, or 8, and the least significant bit is a 1, which corresponds to the zeroth power of 2, or 1.) The output which corresponds to the most significant bit changes to state 1 (or 0) at time $t+1$ if the internal state at time t is greater than or equal to (or less than) 8.

The timing diagram illustrates the temporal evolution of the input states, the internal state, and the states of the LSB and MSB outputs. Note the successive delays between the inputs, the internal state, and the outputs (Fig. 2.2).

The logic circuits of digital electronics can be expressed very simply in terms of automata.

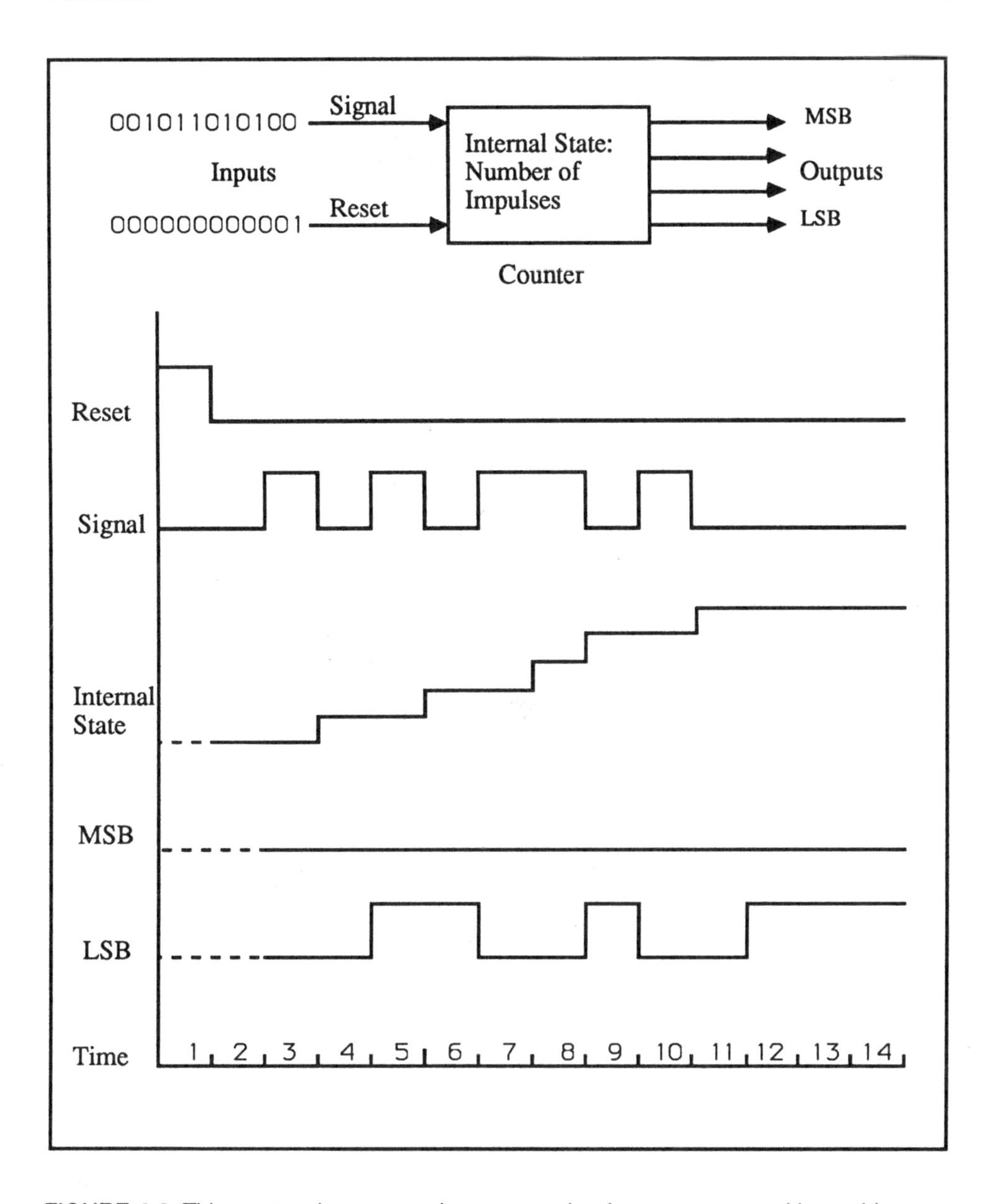

FIGURE 2.2 This 0–15 pulse counter is an example of an automaton with two binary inputs (on the left), one internal state, and four binary outputs (on the right). The timing diagram gives an example of the evolution of the internal state and the outputs for a given sequence of input signals. After having been zeroed by the reset, the internal signal counts the time intervals during which the input signal is a 1. The four outputs code in binary the internal state at the previous time. The MSB output only switches to state 1 if the internal state was greater than 8 at the preceding time, which does not happen during this particular sequence of inputs. The LSB output represents the parity of the internal state: it is in state 1 if the internal state was odd in the previous time interval.

2–1–2 Simplified Definition

In fact, in the examples given in this book, we will simplify the previous definition by considering only automata for which the internal state and the output are the same.

A simplified automaton is defined by its sets of inputs and outputs and by the *transition function*, which gives the output at time $t+1$ as a function of the inputs and sometimes also the internal state (i.e., the output) at time t.

In addition, we will limit ourselves to binary automata, that is to say to two states, for example 0 and 1.

2–1–3 Examples

We will work essentially with two types of automata, boolean automata and threshold automata, which we will define below. Chapter 4 on cellular automata contains other examples of automata.

BOOLEAN AUTOMATA Boolean automata operate on binary variables, that is to say variables which take the values 0 or 1. In logical terms, 0 and 1 correspond to FALSE and TRUE, respectively. The usual logic functions AND, OR, and XOR are examples of transition functions of boolean automata with two inputs. A boolean automaton with k inputs, or of *connectivity* k, is defined by a truth table which gives the output state for each one of the 2^k possible inputs. There are 2^{2^k} different truth tables, or automata.

Let $k = 2$. Here are the truth tables of four boolean logic functions with two inputs:

	AND				OR				XOR				NAND			
Inputs	00	01	10	11	00	01	10	11	00	01	10	11	00	01	10	11
Outputs	0	0	0	1	0	1	1	1	0	1	1	0	1	1	1	0

On the input line of the table, we have represented the four possible input configurations by 00, 01, 10, and 11. The four truth tables correspond to the standard definitions of the following logic functions: AND returns a 1 only if its two inputs are 1; OR returns a 1 only if at least one of its inputs is a 1, XOR is 1 only if exactly one of its inputs is a 1, and NAND is the complement of AND. In logical terms, if A and B are two propositions, the proposition (A AND B) is true only if A and B are true.

Table 2.1 gives the truth tables, the decimal codes (for a definition see the appendix), and the names of sixteen boolean functions with two inputs.

TABLE 1 The sixteen boolean functions with two inputs as defined by their truth tables.[1]

Inputs, Outputs,	11 10 01 00	11 10 01 00	11 10 01 00	11 10 01 00
Code Number, &	0 0 0 0	0 0 0 1	0 0 1 0	0 0 1 1
Name	**0** contradiction	**1** NOR	**2**	**3** inverse 2
Inputs, Outputs,	11 10 01 00	11 10 01 00	11 10 01 00	11 10 01 00
Code Number, &	0 1 0 0	0 1 0 1	0 1 1 0	0 1 1 1
Name	**4**	**5** inverse 1	**6** XOR	**7** NAND
Inputs, Outputs,	11 10 01 00	11 10 01 00	11 10 01 00	11 10 01 00
Code Number, &	1 0 0 0	1 0 0 1	1 0 1 0	1 0 1 1
Name	**8** AND	**9** EQUivalence	**10** transfer 1	**11** 2 implies 1
Inputs, Outputs,	11 10 01 00	11 10 01 00	11 10 01 00	11 10 01 00
Code Number, &	1 1 0 0	1 1 0 1	1 1 1 0	1 1 1 1
Name	**12** Transfer 2	**13** 1 implies 2	**14** OR	**15** Tautology

[1] Their names are taken from mathematical logic. The two functions with codes 0 and 15 have outputs which are independent of their inputs. Four functions: 3, 5, 10, and 12 depend only on one of the two inputs, which they transmit or invert. The functions numbered 1, 4, 7, 8, 11, and 13 are so-called forcing functions: when at least one of their inputs is in a certain state, the output does not depend on the state of the other input. For example, it suffices that one of the inputs of the AND function be in the state 0 for the output to be 0. Finally, for two of the sixteen functions, 6 and 9, the output can only be determined if both of the inputs are known.

THRESHOLD AUTOMATA The state change function of a threshold automaton is defined by:

$$x(t) = Y\left[\sum_j T_j x_j(t-1) - \Theta\right] \tag{2.1}$$

where Y is the Heaviside function, equal to 1 if its argument is greater than or equal to 0, and 0 otherwise. The sum is computed over all of the inputs subscripted by j. T_j is the weight of input j. In other words, the automaton takes the value 1 if the weighted sum of the inputs $\sum T_j x_j$ is greater than or equal to the threshold, and 0 otherwise (Fig. 2.3).

The state change function, which is necessarily a boolean function, is determined by the choice of the weights T_j and of the threshold Θ. A thresholding function can therefore be defined either by its connections and its threshold,

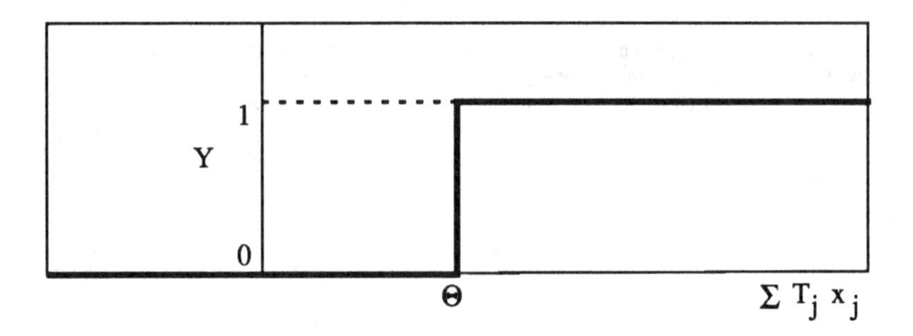

FIGURE 2.3 A thresholding function. The function is 1 if the weighted sum of the inputs $\sum T_j x_j$ is greater than or equal to the threshold Θ, and 0 otherwise.

or by its truth table. If it has many inputs, the definition in terms of the connections and threshold is more compact than the truth table. On the other hand, the truth table speeds up the functional evaluation.

Conversely, not all boolean functions can be put into the form of a thresholding function, as we will see in chapter 7.1.

2–1–4 Automata and Differential Equations

The reader is probably more familiar with the dynamical properties of systems of differential equations, which are used for example to describe physical systems in mechanics or electromagnetism. Automata networks can be considered to be systems of differential equations which have been simplified by extreme discretization. The discretization of time corresponds to the transformation of the differential equation into a finite difference equation. In this manner,

$$\frac{dx}{dt} = f(x, y)$$

becomes

$$x(t+1) = x(t) + f\big(x(t), y(t)\big)$$

The next stage is to replace all of the real variables x and y by binary variables, or automata. The $1 + f$ operator of the finite difference equation then becomes the boolean state change function of the automaton x with two inputs x and y.

The description in terms of automata is a natural one when the components of the system are discrete, as for example atoms and molecules in physics or cells in biology. Sometimes a system of differential equations can be simplified when the continuous variables of the system alternate between different quasi-static states, separated by short transitions. This is the case, for example, of relaxation oscillations observed in highly nonlinear systems of differential equations. One example which we will discuss in this book has to do with cellular biology. The concentration of certain chemical species inside a cell is sometimes negligible, if for example that species is not being synthesized or when its transport across the cell membrane is blocked, and sometimes fixed at saturation when it is present. The concentration of this chemical species can therefore be represented by an automaton, where the state 0 represents the minimum concentration of the product, and state 1 represents its saturation value.

2–2 Automata Networks

2–2–1 Structural Properties

An *automata network* is composed of a set of automata interconnected such that the outputs of some are the inputs of others. It is therefore a directed graph, where the nodes are the automata and the edges are the connections from the output of one automaton to the input of another.

Networks are considered *open* or *closed*, depending on whether or not certain inputs to the automata of the network receive signals from elements outside of the network. External outputs do not make the dynamical properties more complicated, but the same cannot be said for external inputs. For now we will limit ourselves to a description of closed networks, that is to say networks with no external connections.

Figure 2.4 represents the graph of the connections of a network of five boolean automata with two inputs. This graph is equivalent to a set of five logical relations:

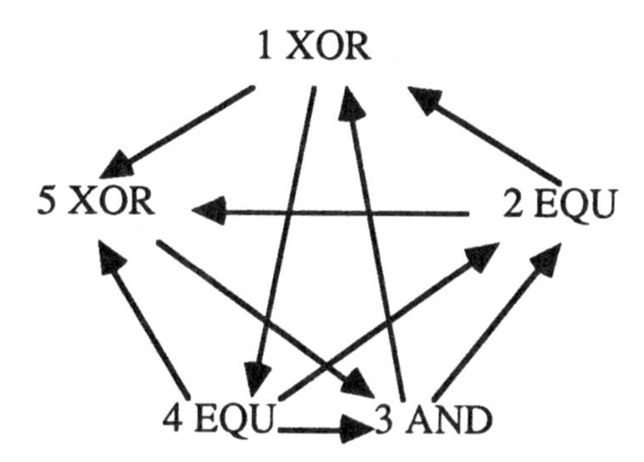

FIGURE 2.4 An network of five boolean automata with two inputs. Each automaton has two inputs and transmits its output signal to two other automata. The XOR and AND functions have been previously defined. The EQU(ivalence) function is the complement of the XOR function—it is 0 only if exactly one input is a 1.

$$
\begin{aligned}
\mathrm{e}(1) &= \mathrm{XOR}(\mathrm{e}(2), \mathrm{e}(3)) \\
\mathrm{e}(2) &= \mathrm{EQU}(\mathrm{e}(3), \mathrm{e}(4)) \\
\mathrm{e}(3) &= \mathrm{AND}(\mathrm{e}(4), \mathrm{e}(5)) \\
\mathrm{e}(4) &= \mathrm{EQU}(\mathrm{e}(5), \mathrm{e}(1)) \\
\mathrm{e}(5) &= \mathrm{XOR}(\mathrm{e}(1), \mathrm{e}(2))
\end{aligned}
$$

where $e(i)$ is the state of the ith automaton.

CONNECTIVITY STRUCTURES This brings us to consider different *connectivity structures*. The connectivity can be randomly chosen; in this case, the definition of the inputs is an exhaustive list of the inputs of each automaton. In this case, we speak of *random connectivity*. We will also consider the case in which each automaton is connected to every other automaton. In this case we say that the network has *complete connectivity*. Finally, *cellular connectivity* is interesting both from a theoretical and a practical point of view. In this case, the automata are distributed on a lattice with small dimensionality (1, 2, or 3) and the connections are between nearest neighbors (see the two following chapters on cellular automata).

2–2–2 Dynamical Properties

ITERATION MODE The dynamics of an automata network are completely defined by its connection graph, the transition functions of the automata, and by the choice of an *iteration mode*. It must be stated whether the automata change their state simultaneously or sequentially, and in what order. We will forego for now a complete description of all of the possible iteration rules, and describe only *parallel iteration*. In this mode, all of the automata simultaneously change their state as a function of the states of the input automata in the previous time step:

$$
\begin{aligned}
&x_1(t+1) = f_1\left(x_1(t), x_2(t), \ldots, x_N(t)\right) \\
&x_2(t+1) = f_2\left(x_1(t), x_2(t), \ldots, x_N(t)\right) \\
&\vdots \\
&x_N(t+1) f_N = \left(x_1(t), x_2(t), \ldots, x_N(t)\right)
\end{aligned}
$$

where $f_1, f_2, \ldots, f_N$ are the transition functions of the automata numbered $1, 2, \ldots, N$. Conversely, in the case of *sequential iteration*, or iteration in series, only one automaton at a time changes its state. Sequential iteration is therefore defined by the order in which the automata are to be updated. Here is an example of sequential iteration:

$$x_1(t+1) = f_1\big(x_1(t), x_2(t), \ldots, x_N(t)\big)$$
$$x_2(t+2) = f_2\big(x_1(t+1), x_2(t), \ldots, x_N(t)\big)$$
$$\vdots$$
$$x_N(t+N) f_N = \big(x_1(t+1), x_2(t+2), \ldots, x_N(t)\big)$$

In the case of parallel iteration, the state following the 00000 state in the network shown in figure 2.4 is 01010 (the convention used is that the leftmost state corresponds to the automaton with the highest number, just as in decimal notation the leftmost digit represents the highest power of ten). On the other hand, in the case of iteration in series as defined above, the state following 00000 is 11010 after the

TABLE 2 Successor table[1]

code	initial state	code	successor	code	initial state	code	successor
0	00000	10	01010	16	10000	2	00010
1	00001	18	10010	17	10001	26	11010
2	00010	27	11011	18	10010	19	10011
3	00011	3	00011	19	10011	11	01011
4	00100	9	01001	20	10100	1	00001
5	00101	17	10001	21	10101	25	11001
6	00110	24	11000	22	10110	16	10000
7	00111	0	00000	23	10111	8	01000
8	01000	8	01000	24	11000	4	00100
9	01001	16	10000	25	11001	28	11100
10	01010	25	11001	26	11010	21	10101
11	01011	1	00001	27	11011	13	01101
12	01100	11	01011	28	11100	7	00111
13	01101	19	10011	29	11101	31	11111
14	01110	26	11010	30	11110	22	10110
15	01111	2	00010	31	11111	14	01110

[1] On the right side of each half-table are the binary configurations and corresponding decimal codes of the successors of 32 initial configurations, shown on the left, for the parallel iteration of the network shown in figure 2.4.

fifth time step, when all of the automata have been updated exactly once. The difference between the two output states is due to the fact that in sequential iteration, the updating of the 5th automaton uses the already modified states of automata 1 and 2.

In the discussion that follows, as well as in chapters 3 and 4, we will talk only of *parallel iteration.*

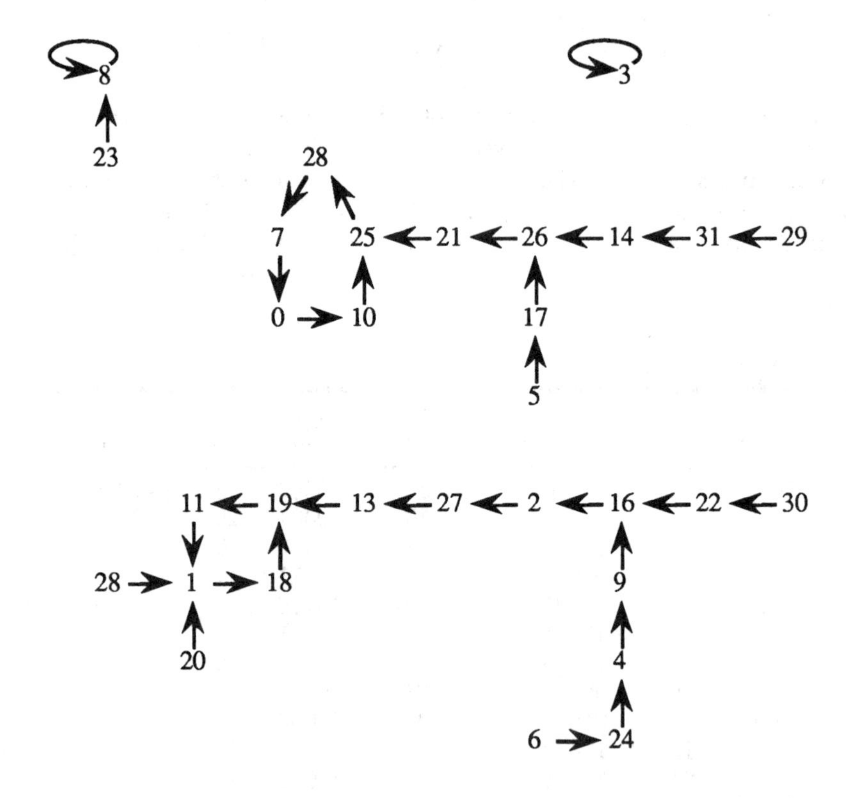

FIGURE 2.5 Iteration graph of the network of figure 2.4 (see table 2.2). The numbers from 0 to 31 refer to the decimal representations of the 32 binary configurations of the network. The most significant bit corresponds to the state of the 5th automaton; the least significant bit corresponds to the state of the first automaton. The arrows show the temporal order of the configurations. Note that there are four different basins of attraction. State number 3 is an isolated fixed point. State number 8 is another fixed point. The other, larger, basins are composed of the configurations which converge toward the limit cycles with periods 4 and 5.

ITERATION GRAPH There are 2^N possible configurations for a network of N boolean automata. We go from one configuration to the next by applying the state change rule to each automaton. The dynamics can therefore be completely described by a successor table like the one shown in table 2.2. The input column of the table lists all 2^N possible configurations, taken as initial conditions, and the output column lists the configurations at the following time step.

From this table, we can draw a directed graph, the *iteration graph*, where the nodes are the configurations of the network and the directed edges indicate the direction of the transitions of the network from its configuration at time t to a new configuration at time $t+1$.

Figure 2.5 represents the iteration graph of the previous network for the case of parallel iteration. This graph contains the $2^5 = 32$ possible states. It illustrates the fundamental dynamical characteristics which we will define below. It is analogous to the trajectories in the dynamics of continuous systems (Fig. 2.6). Note the analogies between the different attractors, fixed points, and limit cycles.

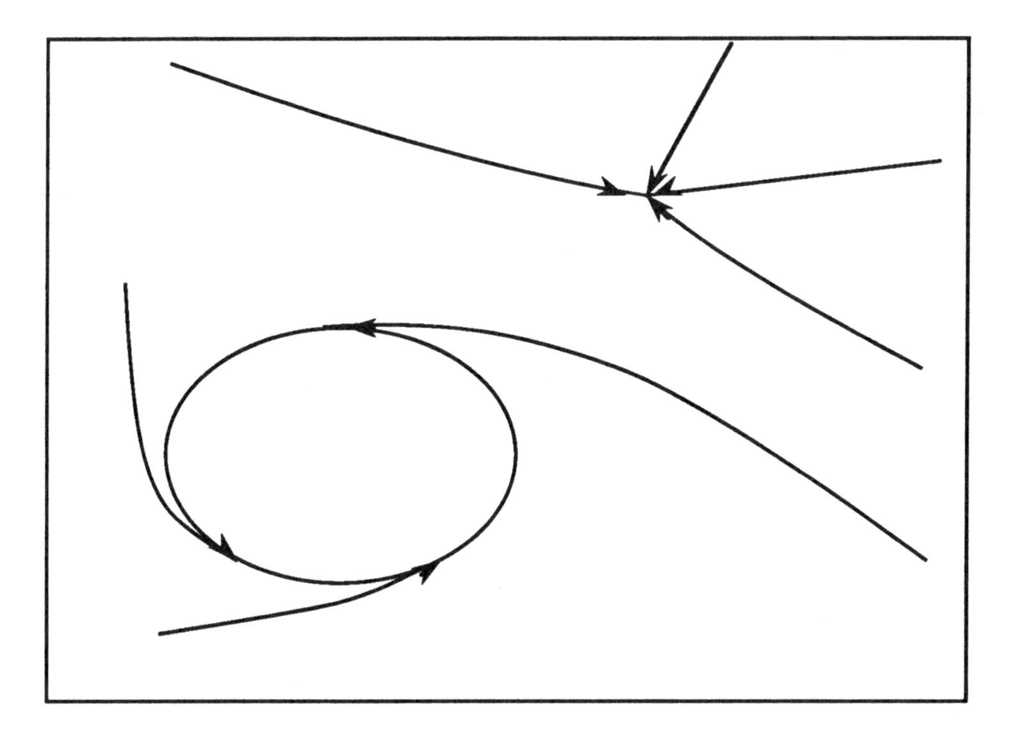

FIGURE 2.6 Trajectories of a continuous dynamical system. There are two basins of attraction; one is a limit cycle and the other is a fixed point.

ATTRACTORS Since an automata network is a deterministic system, if the network reaches a state for the second time, it will go through the same sequence of states after the second time as it did the first. Therefore, the system will go into an infinite loop in state space. These loops are called the *attractors* of the dynamical system, and the time it takes to go around the loop is called the *period* of the attractor. If this period is 1, as is the case for the configuration numbered 8 in the example shown above, the attractor is a *fixed point*. We speak of *limit cycles* if the period is greater than 1. The set of configurations which converge toward an attractor constitutes a *basin of attraction*. The network shown in the above example has four attractors.

Clearly it is only possible to construct a complete iteration graph for small networks. For the large networks used for the applications described in this book, we must be content to describe the dynamics of the system by characterizing its attractors.

In this way we can try to determine:

- the number of different attractors,
- their periods,
- the sizes of the basins of attraction (the number of configurations which converge toward each attractor), and
- the duration of the *transient* (that is to say the time of evolution from the Garden of Eden—the initial configuration—to the attractor).

The notion of *distance* is also very important. The *Hamming distance* between any two configurations is the number of automata which are in different states.

Computer simulation is of course an ideal tool for such studies. In fact it is the ease with which simulations can be carried out on automata networks compared to systems of differential equations which motivates the choice of the former as a mathematical model. We will also see that the theory can sometimes be used to make predictions about the dynamical properties of a network with a given structure, as in the simple case which we will now describe.

2–2–3 A Simple Case: The "Crabs"

Consider a network like the one shown in figure 2.7 in which automata have only one input, but can have many identical outputs.

There are four possible boolean automata with one input.

Two are fixed at state 0 or 1, regardless of the input. We will not discuss these further. (These two invariant automata are always among the 2^{2^k} possible boolean automata with k inputs.)

One of the remaining two automata simply transmits the input signal—we will refer to it as $+$. The other inverts the signal—we will refer to it as $-$.

By connecting together automata with one input to form a network, as shown in figure 2.7, we obtain independent subnetworks which look like crabs. Independent

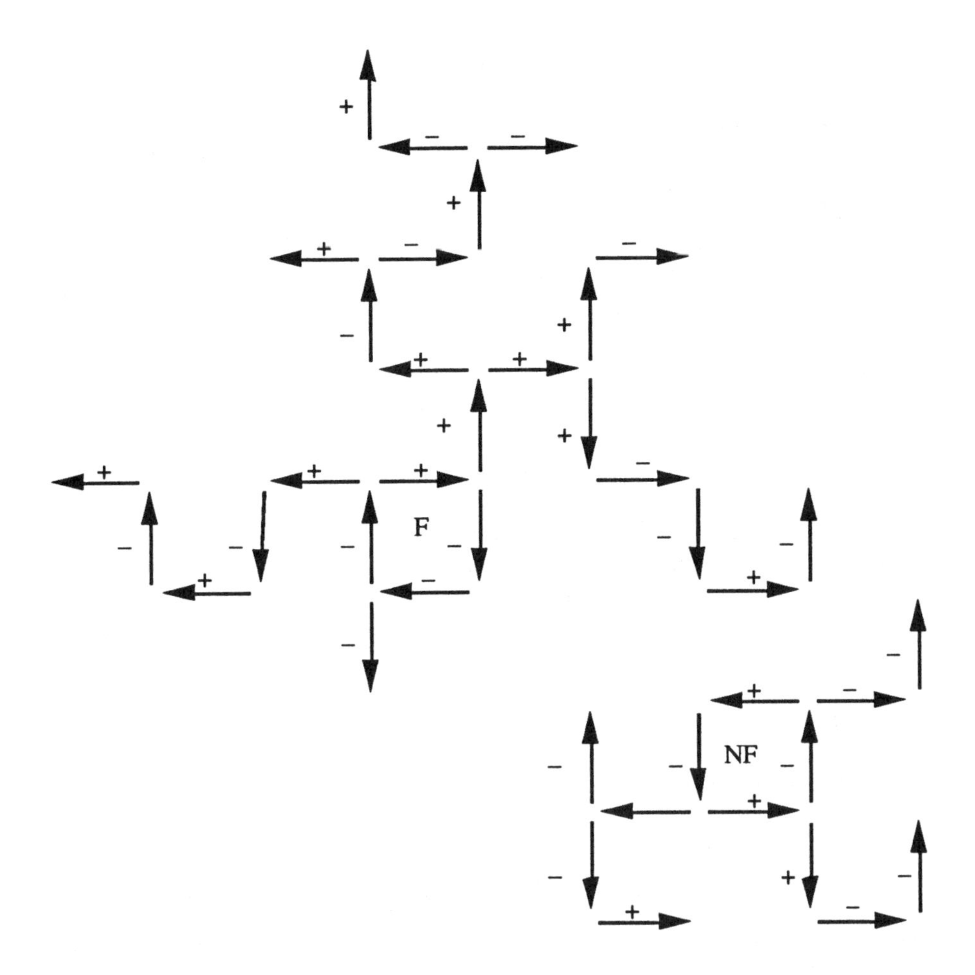

FIGURE 2.7 Network of boolean automata with one input with a two-crab structure. F indicates a frustrated loop, and NF a loop which is not frustrated. The arrows indicate the input connections and the + and − symbols the type of automaton which is at the head of the arrow.

subnetworks can appear in networks with higher connectivity, but they have a higher probability of occurring in networks of automata with one input.

These subnetworks have a particular structure. Each subnetwork contains a loop and branches which radiate outward, carrying the signal from the loop. The input signal cannot enter a loop, because in that case the node by which it would enter would have two inputs: one internal to the loop and the other external. For this same reason, a subnetwork can have only one loop, for the direction of propagation along a branch is defined by the loop from which it stems; two loops cannot therefore be connected by a branch. We will use the image of a "crab" to describe the subnetworks, where the "heads" are the loops and the "legs" are the branches stemming from the heads.

The dynamics of the states of a leg is the propagation of a signal from the input of the first automaton to the last automaton of the leg. At each step, the signal is either transmitted unchanged, or inverted, depending on the function of the automaton. At time t, the state of the ith automaton of the leg depends only on the state of the first automaton of the leg at time $t - i$. If we start from an arbitrary initial state, the transitory phase lasts only as long as the signals from the initial state are propagating down the legs. In the stable state, the signal which propagates down the legs is the signal from the head of the crab. The maximum length of the transitory phase is the length of the longest leg in the crab.

Subsequently, everything is determined by the loop of the head. It follows that the period of the crab is that of its head. There are two possible cases, depending on whether or not the crab is frustrated. The notion of frustration, more serious than our zoological tale would imply, was introduced by F. Harari in models of psychological relationships between people and by G. Toulouse in the theory of spin glasses, which we will discuss later. The *frustration* of a loop is defined by the parity of the number of inversions in the loop. If this number is odd, the loop is said to be frustrated; it is not frustrated in the opposite case. In the case of an unfrustrated loop, the signal which has gone around the loop returns unchanged to its starting point. The period of the loop is therefore equal to m, the number of automata in this loop, or to a factor of m, in the case where the initial configuration has certain symmetries. In particular, the period is equal to 1 if the initial state of each automaton is the output corresponding to the state of its input automaton. It follows that an unfrustrated crab has two fixed points.

On the other hand, in the case of a frustrated loop, the signal returns to its initial state only after having gone around the loop twice. The period of the loop is therefore $2m$ for configurations with no symmetries, and an odd factor of $2m$ in certain special cases. An important point is that a frustrated loop has no fixed point.

The dynamics of boolean networks with one input can be characterized in this manner by a theoretical study. We can determine the transitory length, the period of the attractors (the smallest common multiple of the periods of independent crabs, these periods being determined by the length and the frustration of the loops), and the number of attractors (the product of the number of attractors of each crab—for an unfrustrated crab this number is of the order of $2^m/m$, or $2^m/2m$ for a frustrated crab if the symmetries are ignored). The sizes of the basins of attraction are of the same order for all of the attractors—it is the product of the period by 2^p, where p is the total number of automata in the legs. Therefore there are no isolated fixed points.

In the example shown in figure 2.7, the reader can verify that there are two crabs, one frustrated and the other not. The unfrustrated crab has two fixed points (stable states), a limit cycle of period 2, and three limit cycles of period 4. The frustrated crab has two limit cycles of period 8. The longest transition lasts seven time intervals. Each basin of attraction has a number of configurations equal to period multiplied by 2^{32}. These considerations allow us to describe the dynamics of a network of 40 automata with a configuration space which includes 2^{40} elements, that is, of the order of $1,000,000,000,000$ configurations.

We will see in later discussions that unfortunately theory can only solve for exceptional cases. Nonetheless, two of the concepts which we have developed for crabs are sometimes applicable: those of independent subnetworks and of frustration. Note that the loop is the simplest structure which requires that the set of configurations of the network be known in order to understand its dynamics. In the case of an open loop, the states of the automata of the chain can be determined after the transitory phase by the state of the first automaton.

THREE
Cellular Automata

3–1 Definitions

Cellular automata are automata distributed on the nodes of a periodic lattice, a discrete geometrical structure invariant under certain translation and rotation operations. The set Z of positive and negative integers is isomorphic to a one-dimensional cellular lattice (i.e., an infinite line of automata). Historically, the first cellular automata proposed by von Neumann were on the nodes of a two-dimensional square grid.

The connectivity—said to be cellular—of each automaton is limited to a certain neighborhood, which usually consists of its nearest neighbors. The structure of the neighborhood preserves the translational and rotational symmetries of the lattice.

Theoretically the lattice is infinite, but in practice it always has edges, which must be taken into account when programming (see the algorithms in the appendix). Several figures in this book represent configurations of cellular lattices; in all of these figures, periodic boundary conditions were assumed. This means that the right edge of the figure is connected to the left edge, and possibly the top to the bottom. In other words, the connectivity structure is circular rather than linear, or toroidal rather than planar (Fig. 3.1).

The state change rules are in principle the same for all of the automata in the lattice. However, certain authors have expressed an interest in cellular structures with different automata, known as inhomogeneous cellular automata. We will study these lattices in chapter 11.

The parallel iteration mode is almost always chosen.

The first studies on automata networks, in particular on von Neumann's self-reproducing automata, dealt with cellular automata. This chapter is devoted to the simplest such structures: one-dimensional cellular automata.

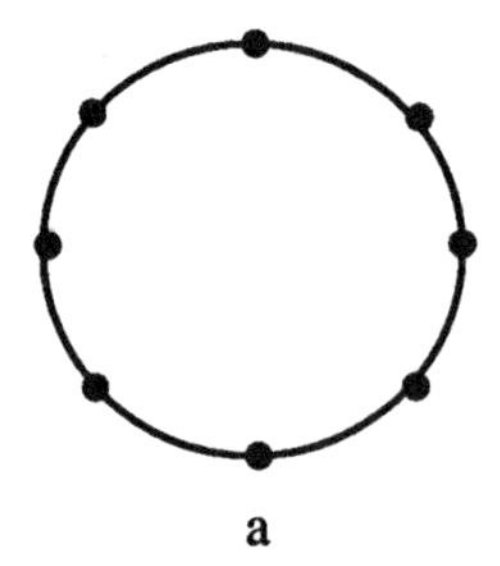

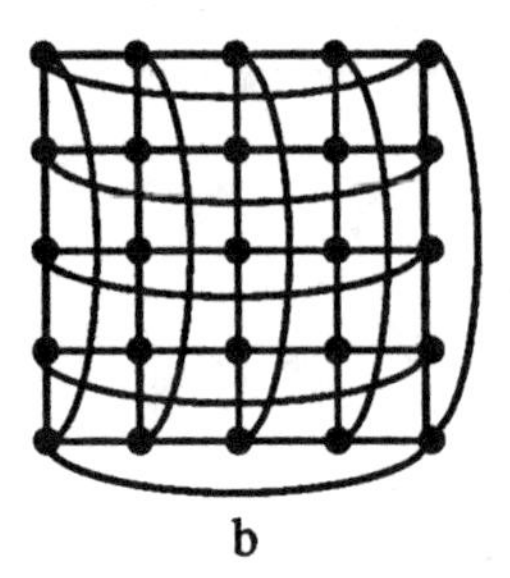

FIGURE 3.1 Periodic connectivity structure of the edges. The automata are represented by black dots and their connections by lines. The automata on the edges of the lattice are connected to each other. a) A linear structure where the automata on the two ends are connected. b) A square lattice, where the connections from one edge to another are represented by curved lines. The automata on the right edge have as neighbors to the right the automata on the left edge, and the automata on the left edge have as neighbors to the left those on the right edge. In the previous sentence, we can replace "left" by "top" and "right" by "bottom."

3–2 One-Dimensional Cellular Automata with Three Inputs

A one-dimensional lattice is but an infinite chain of automata. The simplest neighborhood consists of the automaton itself and its two nearest neighbors. It follows that each automaton has three inputs and computes its new state from its own state and that of its two neighbors. An input state is therefore represented by three bits, the first of which is the state of the neighbor on the left, the second the state of the automaton, and the third the state of the neighbor on the right. Thus there exist $2^3 = 8$ different input states and $2^8 = 256$ different state change rules. These 256 rules, numbered from 0 to 255, encode in decimal notation the eight output bits corresponding to the eight input configurations (from 000 for the least significant bit to 111 for the most significant).

For example, rule 90 (01011010 in binary notation) has the following truth table:

Input states	111	110	101	100	011	010	001	000
Outputs	0	1	0	1	1	0	1	0

It returns the value of 1 for the inputs 001, 011, 100, 110, and 0 for all other inputs.

One simple way to picture the dynamics of these lattices is to use a representation in which each row represents the state of the lattice at time t, and where time increases from one row to the next. The following figures (3.2) show the evolution in time of several cellular lattices using this representation, which was proposed by

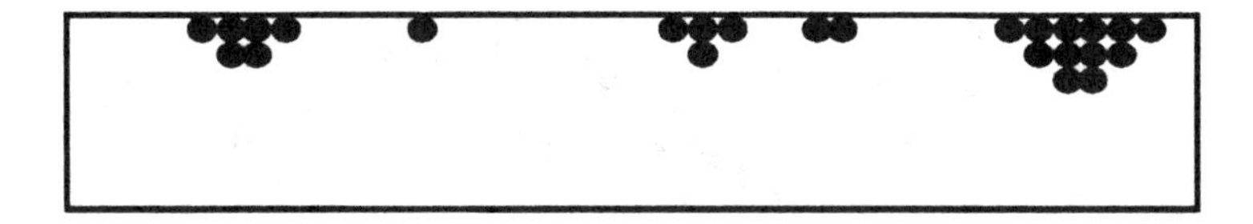

Function 128 (10000000)

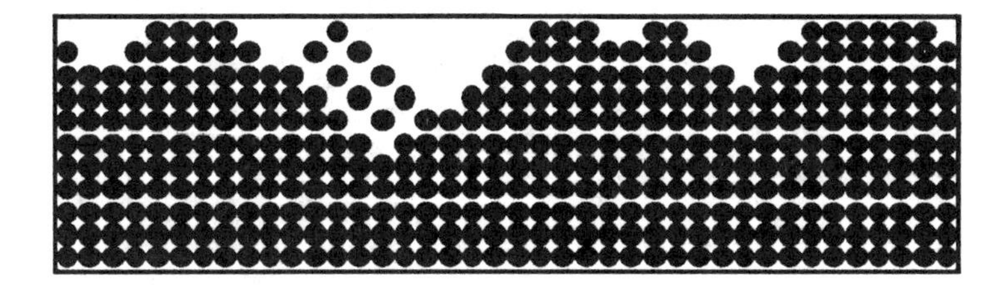

Function 250 (11111010)

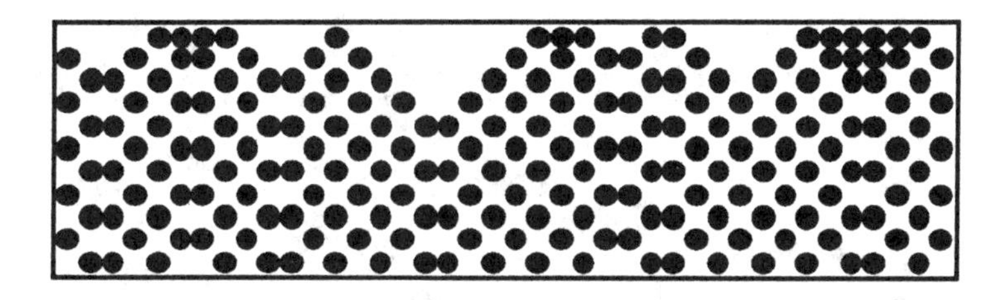

Function 178 (10110010)

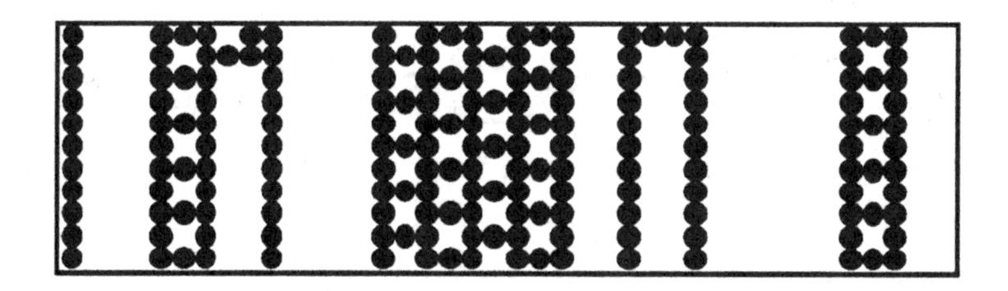

Function 108 (01101100)

FIGURE 3.2 Iterations of linear lattices of cellular automata with three inputs. (cont'd.)

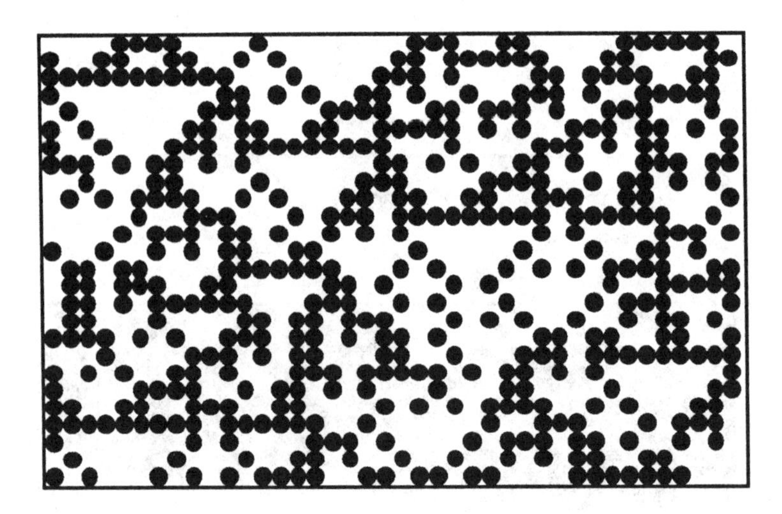

Function 90 (001011010)

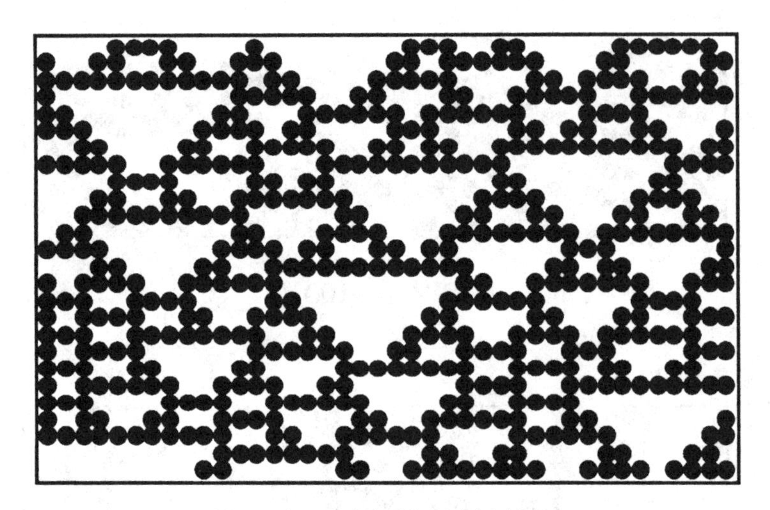

Function 126 (01111110)

FIGURE 3.2 (cont'd.) Automata in state 1 are represented by black dots; those in state 0 are invisible. Each row represents the state of the lattice at time t, and the following row the state at time $t + 1$. Time increases from top to bottom. In all the cases except one, the same initial state is iterated using a different rule, as defined by its decimal code (the binary representation of the truth table is shown in parentheses). Only rule 108 starts with a different initial configuration.

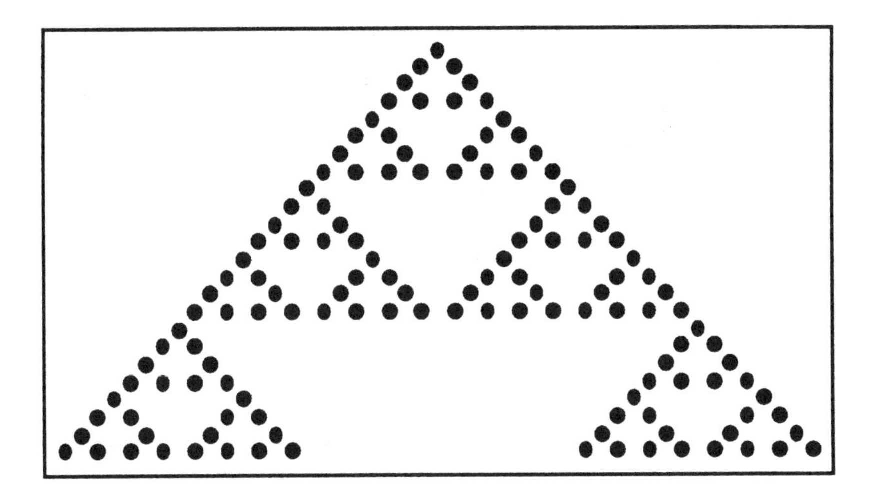

Function 90

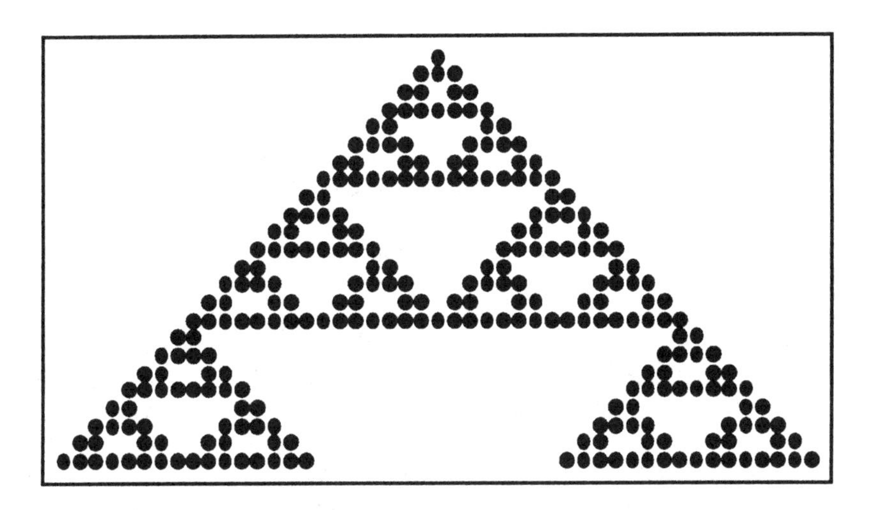

Function 126

FIGURE 3.3 Iteration of rules 90 and 126 from an initial configuration with a single automaton in state 1.

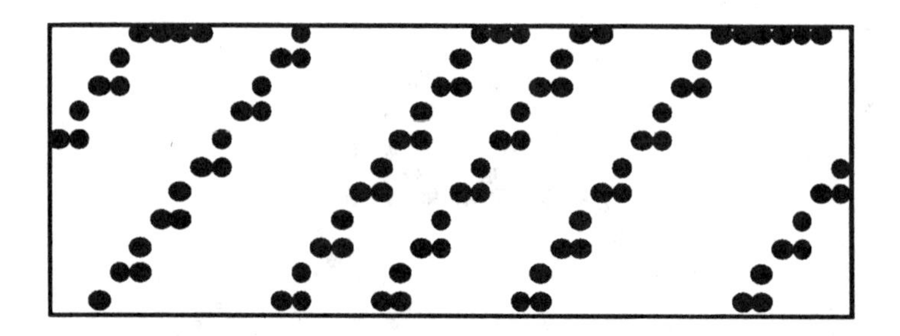

Function 2 (00010100)

FIGURE 3.4 Rule 20 is not symmetric with respect to left and right inputs. The attractor, which was reached after a single iteration, has period 2, but with a translation toward the left.

S. Wolfram. The initial configuration represented on the first row is chosen randomly, and each automaton has an equal probability of starting in the state 0 or 1. In each one of these cases, the state change rule is referred to by its decimal code.

Three types of behaviors emerge from these figures, from which we can define three classes of functions.

1. Certain boolean functions have such "strong" attractors that all configurations converge rapidly toward an *invariant homogeneous state*, composed of all 0's (as in rule 128) or all 1's (as in rule 250). Specifically, for example, in the case of rule 128, with the exception of the configuration of all 1's which, being invariant, constitutes an isolated fixed point, all configurations with at least one 0 converge toward the attractor containing only 0's.
2. Again we see the classical case of *short-period* attractors, which have periods of 1 or 2, depending on the initial configurations (rules 108 and 178). The number of different attractors is very large—it grows exponentially with the number of automata.
3. Finally, a relatively new situation is that of *long periods*, too long to be easily observable (rules 90 and 126). These periods grow exponentially with the number of automata in the lattice. The concept of an attractor loses its meaning somewhat in this case.

This notion of a long period can in fact be defined with some precision. If the average period of a lattice of size N is of the order of a power of N, the period is considered short. The period is considered long if it is exponential in N.

The spatial structure illustrated by figure 3.2 in the case of the long periods corresponding to rules 90 and 126 is also noteworthy. We can see that it has a certain regularity in "texture." The triangles which point downwards can be seen in several places. In fact, a statistical study of this figure would reveal fractal structures, which can be seen directly by starting with particular initial conditions, such as a configuration in which a single automaton is in state 1 (Fig. 3.3).

Symmetric boolean functions with three inputs, that is to say functions for which the inputs on the right and the left behave symmetrically, belong to one of these three classes. The other functions also have the same type of behavior, with the addition of a translation to the left or to the right (see the example of rule 20 in figure 3.4).

We could of course consider larger input neighborhoods, such as five neighbors, or we could work with more complicated automata with several states. The observed dynamical behavior always belongs to one of the three classes defined above. S. Wolfram's book, *Theory and Applications of Cellular Automata* (World Scientific, 1986), gives many examples of cellular automata with larger neighborhoods and more internal states. The possible existence of even richer behaviors, known as "universal computers," is still an open question, and we will come back to this notion when we discuss two-dimensional cellular automata.

3–3 One-Dimensional Cellular Automata with Two Inputs

A special case of automata with three inputs is the case in which the state of the automaton at time t is not used to determine its state at the following time. From a functional point of view, then, the automaton has only two inputs. (More generally, an automaton with n inputs can always be considered a special case of an automaton with a greater number of inputs.) Again, we find that these automata behave in the same way as the automata with three inputs—the behaviors fall into the three preceding classes.

Note that if there is an even number of automata, the lattice is in fact composed of two functionally distinct sublattices. Let's number the automata. At time t, the states of the even- (resp. odd-) numbered automata depend only on the states of the odd- (resp. even-) numbered automata at the previous time. At the next timestep, it is these same even-numbered automata which will determine the states of the odd-numbered automata (Fig. 3.5).

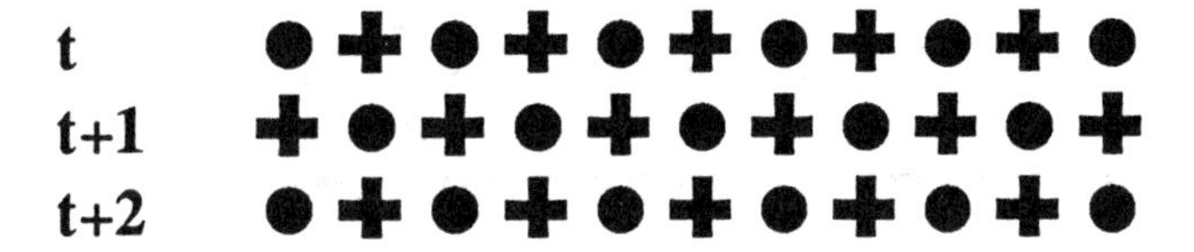

FIGURE 3.5 A linear lattice of cellular automata with two inputs can be separated into independent sublattices, the nodes of which are represented here by dots or crosses.

Consequently, only automata with the same parity interact at any given moment, and the lattice can be decomposed into two independent sublattices. It is convenient, then, to start from an initial configuration of all 0's for one of the sublattices and to limit thc analysis to the other one. This is what is done in figures 3.6 and 3.7.

Since there are only 16 boolean functions with two inputs, it does not take long to study all of them completely. Moreover, the complement of each function can be obtained by inverting the 0 and 1 outputs for each of the input configurations. The two functions have isomorphic iteration graphs, since each configuration of one graph is the complement of the configuration of the other. Therefore, thanks to the 0/1 symmetry, we can limit our study to the eight functions of table 3.1 with decimal codes 0 to 7.

The same initial configuration is iterated in figure 3.6 for 7 different boolean rules, giving attractors with period 1.

Two of these eight functions, with codes 3 and 5, are in fact functions of one input that only transmit either their left or their right input. Therefore, all initial configurations are preserved, although they undergo a translation.

The function with code 0 always converges in one iteration to the configuration in which all automata are in state 0, regardless of the initial state.

The two functions with codes 2 and 4 have analogous roles: they detect differences in the states of the neighboring units. Function 2 detects the decrement from left to right, 10, while the other detects the increment 01. Given any initial configuration, they converge in one iteration to a configuration which is then preserved, though translated, in subsequent iterations.

TABLE 3.1 Eight boolean functions with two inputs. The eight other functions are their complements.

Code	LR	LR	LR	LR	Left Input, Right Input
	00	01	10	11	Input Configuration
0	0	0	0	0	"contradiction"
1	0	0	0	1	"AND"
2	0	0	1	0	unsymmetrical
3	0	0	1	1	transmits left input
4	0	1	0	0	unsymmetrical
5	0	1	0	1	transmits right input
6	0	1	1	0	"XOR", exclusive or
7	0	1	1	1	"OR"

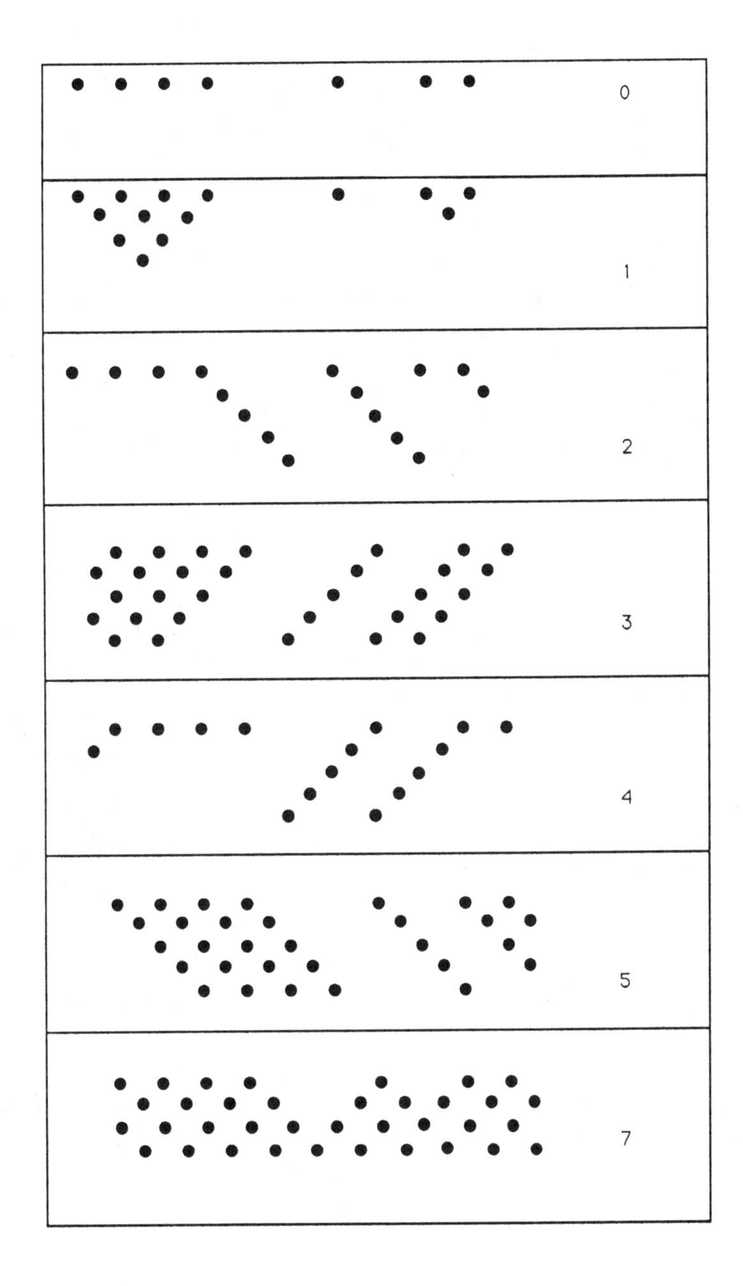

FIGURE 3.6 Iterations of linear lattices of cellular automata with two inputs.

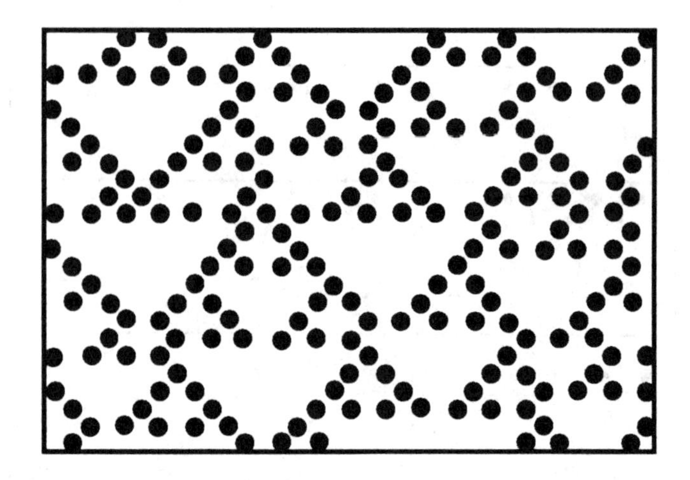

FIGURE 3.7 Iteration of the XOR function.

The AND and OR functions, which have code numbers 1 and 7, also have a simple attractor: all 0's for AND, and all 1's for OR. Homogeneous regions grow out evenly from initial regions of the same type.

Only the XOR function, with code 6, and its complement the EQUivalence function, with code 9, yield interesting behaviors with very long periods (Fig. 3.7).

One-dimensional cellular automata lattices are textbook models which, despite very simple structures, exhibit a variety of behaviors. It is very easy to program their dynamics (on a micro-computer, for example) and to study their evolution.

References

QA 267.5. C45. T481 (Physics)

S. Wolfram's book, *Theory and Applications of Cellular Automata* (World Scientific, 1986), contains a great many studies and figures on the dynamics of one-dimensional cellular automata. Some algebraic methods are described which can exactly predict the dynamical evolution of a lattice for certain transition functions.

FOUR

Two-Dimensional Cellular Automata

Two-dimensional cellular automata have a relatively ancient history, from von Neumann's self-reproducing automata of the 1940's to Conway's "game of life" (1960). These systems have proven to be valuable modeling tools in the fields of the physics of growth, physical chemistry, and hydrodynamics, for they make it possible to follow the mechanisms of interaction between automata and to observe the state of the lattice at any given moment in the iteration. In these fields the dynamical properties are often described by partial differential equations with nonlinear terms which are not solvable by analytical methods. Classical direct methods are computationally intensive and require sophisticated programming techniques (mesh problems). Cellular automata, when they can be used, are invaluable because of the speed of the computations, the simplicity of the programming involved, and the ability to directly visualize the evolution of "patterns" generated by these systems (interfaces, crystals, hydrodynamic structures, flame fronts, etc.).

Finally, the potential applications in VLSI technology are very promising, since the only electronic structures which have been built to date are two-dimensional. Impurities, insulating oxides, and metallic contacts which enable the different electronic functions of the chip to be implemented are deposited directly on the surface of the silicon. Therefore it is easy to design a lattice of cellular automata on a VLSI circuit. Dedicated structures, or structures which are specialized for a specific computation, have already been built. They are known as systolic arrays and are used for matrix computations, especially parallel matrix multiplication.

In two dimensions, the connectivity structure is related to the symmetries of the lattice, which have already been categorized by crystallographers. As long as we require that the lattice have both rotational and translational symmetry about each node, the rotational symmetry can only be of order 2, 3, 4, or 6. Therefore we study lattices which are triangular (each automaton has six nearest neighbors), square (four nearest neighbors), or hexagonal (three nearest neighbors) (Fig. 4.1).

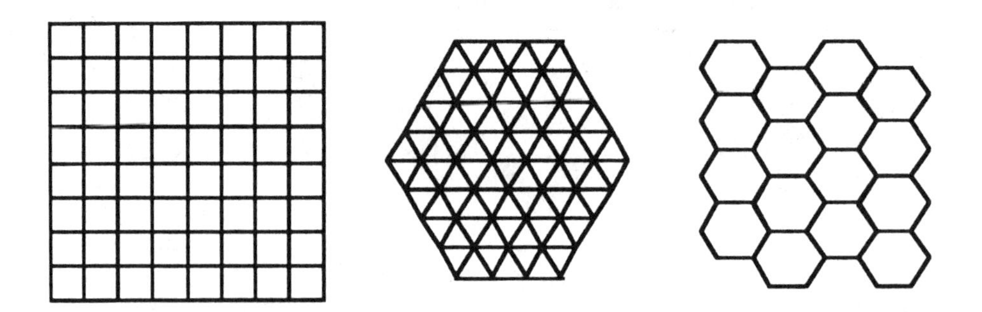

FIGURE 4.1 Square lattice (left), triangular lattice (center), and hexagonal lattice (right).

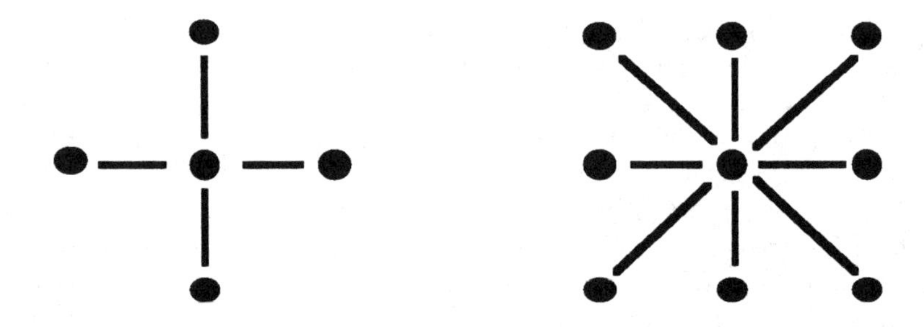

FIGURE 4.2 Von Neumann neighborhood (left) and Moore neighborhood (right).

We will mainly refer to lattices with square grids, or square lattices. Two types of connectivities of varying ranges will be discussed. In the von Neumann neighborhood, each automaton has five inputs, consisting of itself and its four neighbors. In the Moore neighborhood, it has nine inputs, consisting of itself and its eight nearest neighbors (Fig. 4.2).

There is an abundance of literature on the properties of these lattices. We will only discuss a simple model of crystal growth based on counter automata, and an application of cellular automata to hydrodynamics. We will also briefly discuss Conway's "game of life." For more variety, the reader can consult S. Wolfram's impressive compilation *Theory and Applications of Cellular Automata* (World Scientific, 1986).

4–1 Counters and Growth

With such large neighborhoods ($k = 5$ or 9), a complete study of the set of boolean automata is not feasible. As in the case of one-dimensional cellular automata, we are interested in transition rules that conserve the local symmetries of the lattice. This is true of the rules for counters, where the transition function depends only on the sum of the states of the neighborhood.

For *threshold automata*, we define a threshold which is a real number or an integer between -1 and $k+1$, where k is the input connectivity of the automaton. The sum of the states of the neighbors is evaluated and compared to the threshold; the new state of the automaton is 1 if this sum is greater than or equal to the threshold, or 0 otherwise.

In a cellular lattice, weak thresholds (negative or close to 0) favor the growth of zones of automata in state 1, whereas strong thresholds (close to k), favor the growth of zones of automata in state 0.

To illustrate this phenomenon, we can follow the evolution of the dynamical behavior of a lattice as the threshold is increased. In this example we will use a square lattice with a von Neumann neighborhood (5 inputs), but all of the results can easily be extended to the other cases discussed. For negative thresholds, in a single iteration the entire lattice is in state 1. If the threshold is positive but less than or equal to 1, zones of state 1 grow around existing automata in state 1 (the *seeds*). When these seeds are absent, there is no such growth.

The most interesting phenomena occur for intermediate threshold values. For a threshold between 1 and 2, the condition for growth is that at least two neighbors must be in state 1. Groups of 1's are still favored, but isolated 1's are destroyed. The seeds must consist of at least three unaligned neighboring automata in state 1. Growth only occurs at the *edges* of *facets* (Fig. 4.3 and 4.4).

If the groups of 1's are far enough apart, the growth stops when the convex envelope of the initial configurations is full of 1's.

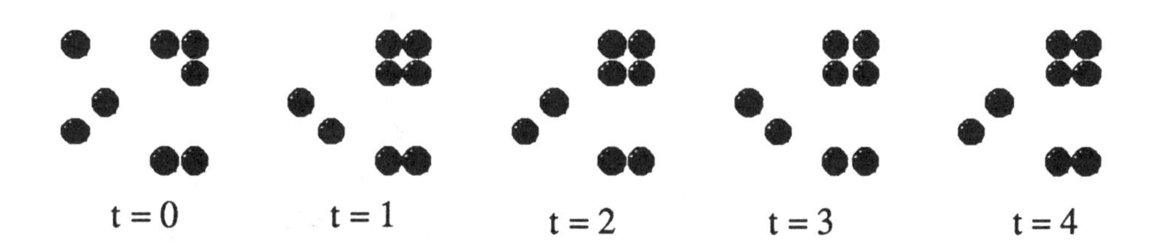

FIGURE 4.3 The evolution of some simple configurations for cellular automata with thresholds of 1.5. The isolated point disappears, the horizontal pair is fixed, and the diagonal pair oscillates with a period of 2. The triplet in the corner evolves in one iteration to a stable square with four automata.

FIGURE 4.4 Growth of the facets of a crystal from one seed (cellular automata with thresholds of 1.5).

This mode of growth is a fairly good representation of the growth of crystals in thermodynamic equilibrium, that is to say when the bath is cooled down slowly enough. This is the case for regular mineral crystals, such as quartz. The convex envelope of the seeds corresponds to the equilibrium shapes. Growth happens on the facets, but from time to time an impurity, an imperfection of the lattice, or even an isolated atom "stuck" on the facet allow the growth of a new row of atoms on that facet. In physical systems, it is these irregularities due to the imperfections of the crystal or to thermal excitation which allow growth from microscopic seeds to beautiful macroscopic crystals.

For a threshold between 2 and 3, there is no growth along the facets, which are instead stripped of all of their isolated atoms. For even bigger thresholds, the roles of the 0's and 1's are reversed, and we observe the growth phases of the zones of 0's in the opposite order.

For other neighborhoods and other lattices, we see similar behaviors, but not necessarily for the same thresholds. The shapes of the crystals can also differ: we see diamond-shaped or hexagonal crystals.

4–2 Window Automata and Dendritic Growth

In nature we see other crystalline forms: snowflakes, for example, are crystals which undergo dendritic growth, which produces lacy shapes. This type of growth occurs when the solid seed is much colder than the solution. In this case, the difficulty in dissipating the heat produced by the crystallization leads to the growth of dendrites which reach out to find colder zones of the liquid. This type of growth can be modeled by cellular automata: the problem of heat dissipation from the crystallization is taken into account by not allowing the transition toward the state 1 when the number of neighbors in state 1 is too large. We then use "window" automata. The state of such an automaton can change to 1 only if the number of its neighbors in state 1 is neither too big nor too small. In the case of the von Neumann neighborhood, the following rule allows the growth of the dendritic structure shown in figure 4.5:

If an automaton is in state 1, it stays there.

If an automaton is in state 0, it changes to 1 only if one of its neighbors is in state 1.

FIGURE 4.5 Window automata and dendritic growth. This figure was obtained after 13 iterations from the square seed of four automata in the center of the figure. The different shades of the dots correspond to the successive stages of the iteration (the black dots, for example, correspond to the steps 0, 5, and 10).

4–3 Conway's "Game of Life"

The game of life is based on a very simple cellular automaton which exhibits surprisingly diverse dynamical behaviors as a function of different initial configurations. This behavioral richness has made it popular in mathematical games and elementary programs for novice computer users. The motivations of its author, Conway, are similar to those of von Neumann. We will not describe von Neumann's model of a self-reproducing automaton, since it is quite complex. It is an automaton with 29 states, and the builder configuration contains 200,000 automata in the nonquiescent state. On the other hand, Conway's model is much simpler, and uses a Moore neighborhood with nine inputs. The transition rules are as follows:

- An automaton in state 0 switches to state 1 if three of its neighbors are in state 1 ("it is born"). Otherwise, it stays in state 0.
- An automaton in state 1 stays in state 1 if two or three of its neighbors are in state 1. It switches to state 0 in the other cases. (It "dies," either of isolation or of overcrowding.)

Figure 4.6 shows the evolution of some simple configurations. The reader may find it amusing to follow the evolution of other configurations by hand on graph paper or by writing a computer program. In general most configurations rapidly evolve toward a set of simple attractor configurations, such as squares, honeycombs, or flashing lights.

The glider configuration is quite remarkable. After four iterations, it returns to its initial configuration, having undergone a translation. This is one of the basic ingredients needed to construct a real "cellular computer." Before we discuss the basic principles of this construction, let us give its motivations. One way to calibrate the richness of dynamical behavior of a system is to prove that it is formally equivalent to a Turing machine, or universal computer, which has well-established properties. All that is necessary to build a Turing machine is to know how to construct two logic functions, including negation. Our cellular computer is therefore based on the following elements:

- Gliders, which are likened to binary signals that propagate down the diagonals of the lattice. The presence of a glider is interpreted to be the signal 1, and its absence 0.
- The "cannon" configuration ejects gliders at regular intervals (every thirty iterations). It allows us to have access to a signal in state 1.
- The collision of two gliders destroys them both.

Figure 4.7 shows an example of how to construct a logic function.

The combination of two different logic functions (for example AND and NOT) allows the construction of a Turing machine, just as the combination of integrated circuits (inverter and AND) can be used to build a practical implementation of a computer. Nobody, of course, would consider using Conway's automata to build a computer—the aim of this discussion is only to demonstrate the wealth of behaviors observed with the "game of life."

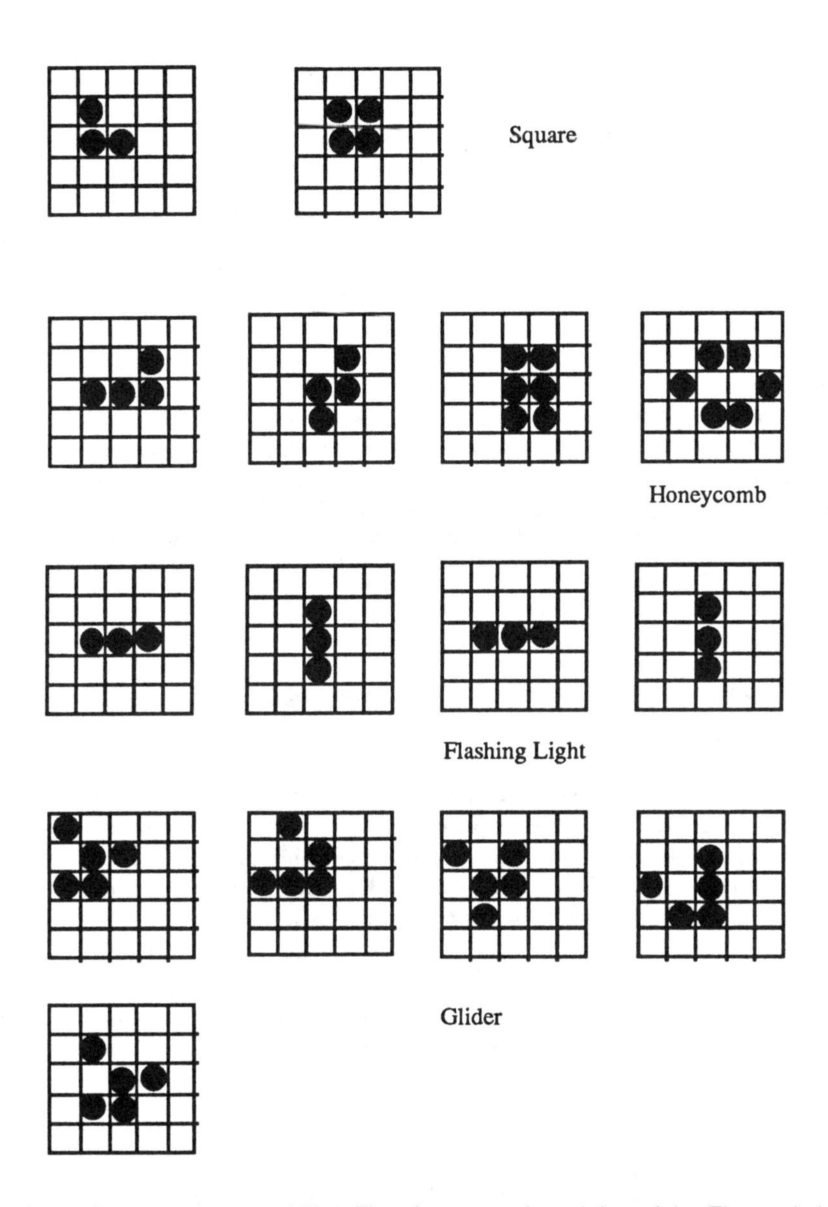

FIGURE 4.6 Conway's "game of life". Time increases from left to right. The evolution of four configurations: the first two configurations evolve toward fixed points, the third toward an attractor of period two, and the fourth is a glider, which reproduces itself after four iterations translated by one square (down and to the right).

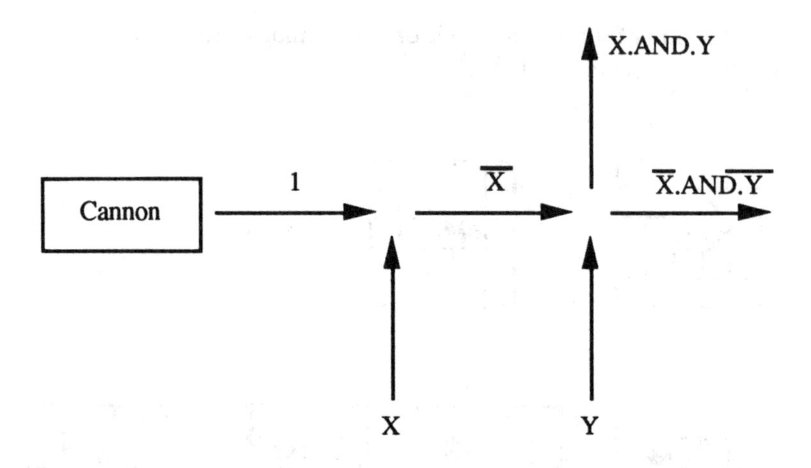

FIGURE 4.7 Configurations of the game of life and logic functions. From the cannon signal which emits gliders every 30 iterations, we can obtain the logic functions NOT $(\overline{X})$, (X), X.AND.Y, and $\overline{X}$.AND.$\overline{Y}$ of the input signals X and Y, represented by the presence or absence of gliders on the propagation axes of the signal, shown here as arrows. By combining elementary logic functions in this way, we can build a universal computer.

4–4 Models of Lattice Gas Dynamics

4–4–1 Principles

In fluid mechanics the behavior of moving fluids is described by a nonlinear partial differential equation, the Navier-Stokes equation. In the case of small velocities, in the linear regime (laminar flow), we can linearize this equation and solve it without much difficulty, analytically if the shapes of the moving objects or the obstacles are simple, and numerically otherwise. If the velocities are large, in the non-stationary regime, instabilities appear and exact analytical methods can no longer be used. Even numerical methods are difficult to use, chiefly because scales of different sizes must be taken into account, which forces grids either to be very small or variable. The observed flow regime depends on Reynold's number R_e, defined as the ratio:

$$R_e = \frac{vL}{\eta/\rho}$$

where v is the average velocity of the fluid, L is a characteristic dimension of the flow, for example the size of an obstacle, and η/ρ is the kinematic viscosity of the medium, the ratio of the viscosity to the specific mass of the fluid. Hydrodynamic instabilities appear when R_e is of the order of ten. The technological importance of these problems (for the design of airplanes or ships, for meteorology, etc.) justifies the huge amount of numerical resources which are currently being deployed to solve them. From the beginning of the computer age, these fluid mechanics problems have been a motivating force behind the efforts to increase computer power.

The Navier-Stokes equation is based on a macroscopic description of physical quantities. It depends on the velocities of the volume elements of the fluid. We can approach the problem from a different, more basic point of view by starting from the microscopic description of the particles of the fluids. First of all, space is discretized and represented by a cellular lattice. The fluid is then represented by particles that move at a constant velocity along the edges of this lattice. Each node of the lattice is an automaton whose inputs are the neighboring nodes. We will work with two-dimensional triangular lattices (the reasons for this choice will become clear later).

The input configuration of the automaton is defined by the presence or absence of a particle coming toward the node on each one of the six edges, plus the possible existence of a particle at rest on the node itself. The state change rule defines the output configuration, that is to say the presence of particles moving from the node toward its neighbors as well as the presence of the stationary particle. In the case of the triangular lattice, 7 bits are sufficient to describe the input configuration, just as 7 bits can describe the output configuration.

The choice of the transition function is not arbitrary: it must obey the conservation laws of mechanics. During a collision, the number of particles, the momentum, and the total energy must all be conserved. Taking into account these laws and the various rotational invariances of the transition function dictated by the symmetries of the lattice considerably limits the choice of functions.

Other constraints must also be taken into account. Fluids are isotropic: this symmetry is broken when we constrain the particles to move along the six directions of the lattice. This restriction at the microscopic level must not lead to a loss of isotropy at the macroscopic level. In addition, we do not want to create "parasitic invariants," macroscopic quantities which are conserved by the rules of propagation along the edges of the lattice but which are not conserved by the Navier-Stokes equation. In fact, it is this last condition which restricts the choice among the two dimensional lattices which are possible *a priori* to triangular lattices exclusively. Finally, the chosen transition function conserves the velocities of the incident particles, except for the following input configurations, shown in figure 4.8:

- head-on collision between two particles;
- head-on collision among three particles;
- collision between two particles, one of which is stationary;
- collision of two particles with a 120° angle between the velocity vectors.

Two of the possible collision rules between particles and the surface of an obstacle are also shown on the same figure. Of course, it is also possible for two particles to meet on an edge, but we do not need to take this case into account, for there is no difference between a simple crossing and a collision with rebound.

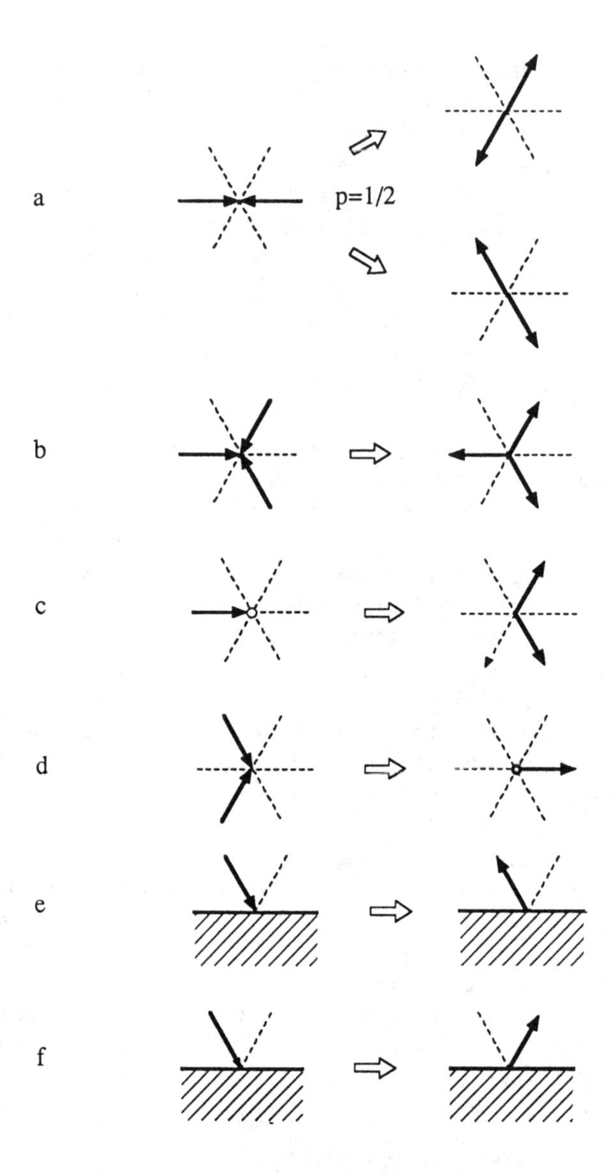

FIGURE 4.8 Collisions on a triangular lattice. The input configurations at time t before the collision are shown on the left, and the output configurations at time $t + 1$ after the collision, are shown on the right. a) A head-on collision: either one of the output configurations shown can be chosen with equal probability. b) A three-particle collision. c) A collision with a stationary particle. d) A two-particle collision at a $120°$ angle. Possible rules for collisions with a stationary obstacle: e) rebound, f) specular reflection. (From D. d'Humières, Y. Pomeau and P. Lallemand, "Images de la physique 1987," *Le Courrier du CNRS*, 89–94.) Reprinted by permission.

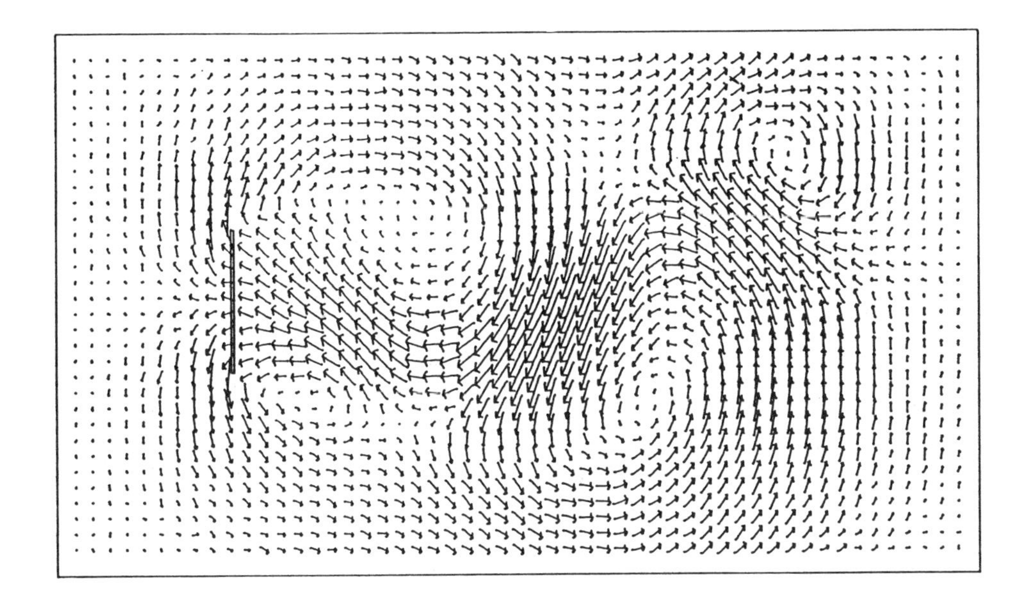

FIGURE 4.9 A von Karman alley behind a thin plate. The result of the simulation by a lattice of cellular automata of the flow of a fluid. (From D. d'Humières, Y. Pomeau and P. Lallemand, "Images de la physique 1987," *ibid.*) Reprinted by permission.

4–4–2 Numerical Simulations

It is possible to test the validity of the model by comparing the results of simulations to theoretical predictions in the cases which can be solved analytically. This has been done successfully in the case of the propagation and attenuation coefficients of planar compression and shear waves.

The most interesting cases are of course those regimes in which it is difficult to solve the Navier-Stokes equation. Figure 4.9 is the result of a simulation obtained after 5000 time steps. It represents a flow near a planar obstacle for a Reynolds number of 150. The lattice contains 512 × 1024 automata. The flow of particles enters from the left and leaves from the right. The boundary conditions for the top and bottom edges are periodic. The arrows correspond to the the local flux averaged over squares of 16 × 16 sites. The average flux of the fluid from left to right has been subtracted to aid visualization. The series of vortices which detach themselves from the plate are analogous to those observed behind an oar which propels a boat forward; it is known as a "von Karman alley."

4-4-3 Perspectives

This model has led to many variations since it was introduced by U. Frisch, B. Hasslacher, and Y. Pomeau in 1985, and since the first simulations by D. d'Humières, Y. Pomeau, and P. Lallemand.

The problem of simulations in three dimensions is much more difficult than in two dimensions. Unfortunately there are no three-dimensional lattices on which interaction rules with no parasitic invariants can be implemented, if all of the particles have the same speed. Therefore it is necessary to go to four dimensions, where such lattices exist, and then to project the results in three-dimensional space. But the large number of neighbors (24) on such a lattice requires truth tables of impressive proportions, which take a long time to compute.

If we stay in two dimensions, many problems of instability in fluids become tractable. Models of flames belong to a class of problems in physical chemistry which require the presence of several types of particles on the lattice. We can add to this class the simulation of multiphase flows in finely divided media and gravitational instabilities, such as the Rayleigh-Taylor problem, where a heavy fluid is initially placed above a lighter fluid.

These models can be computationally intensive, requiring the use of a CRAY, for example, but can easily be parallelized on smaller machines, such as the Connection Machine. A dedicated prototype, the RAP, has already been built for this purpose at the Ecole Normale Supérieure. Its performance for this task is comparable to that of a CRAY1. Several million nodes are processed per second, which allows a direct visualization of the flow (50 images per second).

The use of cellular automata to solve hydrodynamics problems is probably one of the most promising applications of automata networks from a technological point of view: the aim is to build large parallel machines which are specialized for the simulation of fluids moving near obstacles, such as airplane wings. Such machines would play the role of "numerical wind tunnels."

References

In addition to S. Wolfram's book, *Theory and Applications of Cellular Automata*, cited previously, G. Vichniac's article, "Cellular Automata Models of Disorder and Organization," in *Disordered Systems and Biological Organization*, edited by E. Bienenstock, F. Fogelman-Soulié and G. Weisbuch (Springer, 1986), elaborates on the problems mentioned in the first part of this chapter.

The "game of life" is explained in detail by J. Conway in chapter 25 of *Winning Ways*, vol.2, edited by E. Berkelamp, J. Conway, and R. Guy (Academic Press, 1982).

A number of models of lattice gas dynamics as well as the design principles of the RAP can be found in the proceedings of a specialized conference published in *Complex Systems*, vol. 1, no. 4 (1987). math QA 402 .C554

FIVE
The Hopfield Model

This subject has recently captured the fancy of a growing number of physicists of disordered systems. The reasons are simple:

- The model which we will describe resembles another model, that of spin glasses, which is used by physicists to describe the magnetic properties of diluted magnetic alloys. Physicists consider these alloys to be simple physical models of disordered media, such as glasses.
- The existence of a quantity analogous to energy makes it possible to use the full power of the methods of statistical physics to determine the attractors.
- It is possible to generate a network which has predetermined attractors.
- This property allows the network to "learn" configurations which it will subsequently be able to "recover" from an initial condition which is not too far from the attractor previously created.

We therefore have a basic model of learning and recognition in the brain, and in addition, a wiring diagram for an artificial system with the same properties.

The simplicity of this model allows us to derive theoretically its most interesting properties. In this chapter, we will develop some calculations based on a very simple method. We put off until chapter 8 the explanation of results which require the use of more sophisticated methods borrowed from statistical mechanics.

5–1 Definition of the Network

The simplest version of these networks, or "neural networks," as they are somewhat improperly called, is the one proposed by J. Hopfield in 1982.

The basic unit is the threshold automaton, defined in chapter 2. These automata were proposed by McCulloch and Pitts as early as 1943 as models of nerve cells, or neurons. These cells can emit trains of electrical impulses in response to signals from other neurons connected to them.

A simplified way to describe the neuron is to consider it to be an automaton in state 1 when it emits impulse trains and in state 0 when it is in a resting state and does not emit impulses. The inputs of the automaton represent the synapses of the neuron, that is to say the connections with other neurons which influence the firing rate of the given neuron (Fig. 5.1). These synapses can be excitatory, in which case the activity of the neuron is reinforced if the input is in state 1, or inhibitory in the

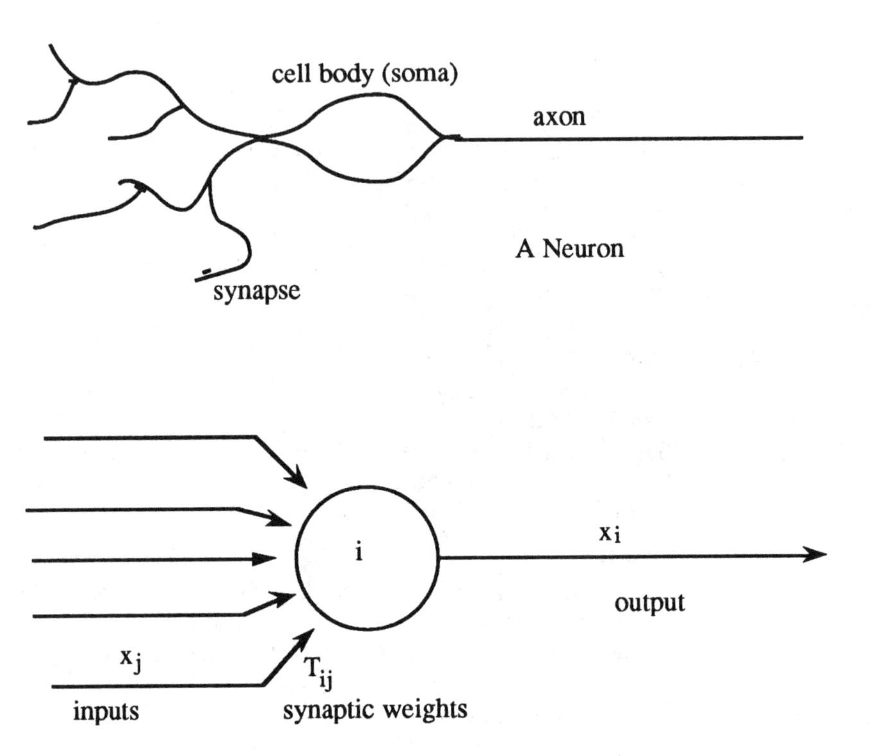

FIGURE 5.1 Schematic drawing of a neuron (top) and its formal representation (bottom). The neuron is excited or inhibited by the signals received on its synapses (on the left side of the figure). These signals are integrated at the level of the cell body, and if the level of excitation is sufficient, a train of impulses is emitted along the axon. At the other end, at the right of the figure, the axonal processes excite other neurons (not shown). The formal neuron i computes the sum of the signals x_j which it receives, weighted by the synaptic weights T_{ij} and emits a signal $x_i = +1$ if this sum is greater than the threshold θ_i.

opposite case. A neuron i is therefore represented by a threshold automaton whose state $x_i(t)$ is updated by the rule:

$$x_i(t) = Y\left[\sum_j T_{ij}x_j(t-1) - \Theta_i\right] \tag{5.1}$$

where Y is the Heaviside function. The sum is computed over all the input automata j (see section 2.1.3). The T_{ij}, called synaptic weights in reference to biology, represent the strengths of the interactions between the input automaton j and the automaton i. Positive synaptic weights are interpreted as excitatory connections and negative weights as inhibitory connections. Θ_i is the threshold of the automaton i. In other words, automaton i is in state 1 if the weighted sum of the states of its inputs $\sum T_{ij}x_j$ is greater than or equal to the threshold, and state 0 otherwise.

In the Hopfield model the connectivity is complete, with the exception of the diagonal terms T_{ii} which are zero. Of course, other non-diagonal terms can also be zero. In addition, the T_{ij} are symmetrical ($T_{ij} = T_{ji}$). Each automaton is thus defined by the synaptic weights of its inputs T_{ij} and its threshold Θ_i.

The random sequential mode of iteration is used: in each time interval, a single randomly chosen automaton is updated. This choice corresponds to the idea that in the brain there exists no internal clock which synchronizes the state transitions of the neurons. This iteration mode, called Monte-Carlo, is also frequently used in numerical simulations of certain physical systems (see section 8.2).

In this chapter we will use a change of variables which will simplify the derivations. We assume:

$$S = 2x - 1 \tag{5.2}$$

which has the effect of making the state variable symmetrical. If $x = 0$, $S = -1$, and if $x = 1$, $S = 1$. This change of variables makes it necessary to change the threshold, replacing it by $2\Theta - \sum T_{ij}$, and to divide by 2 the argument of the Heaviside function, which has no effect on its output. The state change function remains a threshold function. Consequently, if the weighted sum of the S_j (with the same synaptic weights) is greater than the new threshold, S_i takes on the value 1. Otherwise, S_i takes on the value -1.

5–2 Energy and Fixed Points

The symmetrical nature of the connections and the choice of the sequential iteration mode enable us to show the existence of a function which varies monotonically with each iteration. We can then apply a theorem, the Lyapunov theorem, which states that the attractors have a period of 1. The "energy" of a configuration, by analogy with the energy of magnetic systems, is defined by:

$$E = -\frac{1}{2}\sum_{i}\sum_{j}(T_{ij}S_j - 2\Theta_i) \cdot S_i \tag{5.3}$$

When an automaton i changes its state at a given step in the iteration, the change in the energy function is:

$$\Delta E = \sum_{j}(T_{ij}S_j - \Theta_i) \cdot S_i + \sum_{j}(T_{ji}S_j - \Theta_i) \cdot S_i \tag{5.4}$$

In this expression, the states of the automata are those at time t, before the iteration. Automaton i changes its state only if the first term is negative. Since the synaptic weights are symmetrical, the second term is equal to the first, and is therefore also negative. Consequently, at each iteration, either the energy remains unchanged when there are no state changes, or it decreases. However, since the sum is finite, the energy must have a lower bound. It follows that, after some time, the minimum energy is reached, in which case the update of each automaton i leaves its state unchanged. The final configuration is therefore a fixed point of the iteration.

Of course, there is no reason for the fixed points which are reached to be identical—they depend on the initial configuration. Similarly, the energies obtained are not identical. Most of them are only relative minima.

We can thus state the following theorem, attributed to E. Goles-Chacc:

- For iteration in series, the attractors of a symmetrical network of threshold automata with nonnegative diagonal elements are fixed points.

Goles-Chacc has also shown by using analogous arguments for a quantity similar to energy that a parallel iteration of this type of network leads to attractors which are either fixed points or limit cycles of period 2.

All of the assumptions that were made about the network (symmetrical synaptic weights, sequential iteration) are necessary to make the attractors fixed points, except that the diagonal coefficients T_{ii} need not be zero. In fact, if these coefficients are positive, the direction in which the energy varies does not change. Indeed, since ΔE only depends on the off-diagonal elements of the connection matrix, we must add to it $2T_{ii}{S_i}^2$ to make both terms negative when there is a change of state. It follows that ΔE remains negative. On the other hand, if the T_{ii} are negative, the energy function is no longer guaranteed to be monotone. For example, if the coefficient T_{ii} of a single automaton is negative and greater in absolute value than the sum of the absolute values of the corresponding terms of all of the automata to which it is connected, and if in addition its threshold is zero, this automaton will change its state with each iteration.

5–3 The Hebb Rule

For any type of network, we can pose the inverse dynamics problem:

- Given an arbitrary set of configurations, is it possible to construct a network of automata for which this set of configurations is the set of attractors?

In other words, we assume that initially we are given m configurations S^s, called *references* which we would like to be the attractors of a network whose synaptic weights and thresholds must be chosen accordingly (the superscript s refers to the references). The Hebb rule partially answers this type of question. In his book *The Organization of Behavior* (Wiley, 1949), D. Hebb suggested that at the cellular level, learning a cognitive task involved reinforcing synapses of neurons with correlated activity. Translated in terms of automata networks, this is equivalent to "constructing" the network by successive presentations of the reference configurations. A presentation consists of fixing the state of each automaton in its state in the reference. The T_{ij}'s of automata in the same state are then increased, while the others are decreased. The T_{ij}'s are therefore given by:

$$T_{ij} = \sum_{s=1}^{m} S_i^s S_j^s \tag{5.5}$$

This is the *Hebb rule.*

The choice of thresholds is particularly simple if:

- the references are *random.* In other words, if the state of each automaton S_i is chosen randomly with an equal probability of being -1 or 1, independently of the states of the others.
- the references are uncorrelated. The product $S_i{}^s S_i{}^{s'}$ is zero on the average if the reference s is different from the reference s'.

In this case, if we use the variables S, we set the thresholds to zero, which we will assume in the rest of the chapter. It is not very difficult to generalize the results which we will describe when one or two of the hypotheses do not hold.

5–4 Simulation Results

These choices allow us to construct a network which has as its attractors the given configurations, as long as there are not too many (of the order of $0.1N$), they are randomly chosen (and consequently, they have about the same number of -1's and 1's), and finally, the different configurations are uncorrelated. The results of Hopfield's first simulations indicate that for a network of 100 automata, with 15 randomly chosen references, the following facts are very likely to be observed:

1	−1	−1	−1	−1	1	−1	−1
1	−1	−1	−1	−1	1	−1	−1
−1	1	−1	−1	1	−1	−1	−1
−1	−1	1	1	−1	−1	−1	−1
−1	−1	1	1	−1	−1	−1	−1
−1	1	−1	−1	1	−1	−1	−1
1	−1	−1	−1	−1	1	−1	−1
1	−1	−1	−1	−1	1	−1	−1

FIGURE 5.2 Configuration representing the pattern X.

- If we use one of the references as the initial condition, the network evolves toward an attractor which is either this reference or one that is very close to it. It differs from the reference only in the states of a few automata.
- If we start from an initial configuration which is close to the reference, we converge toward the attractor of that reference, in other words toward that reference or toward a nearby configuration.

In order to illustrate these behaviors, we can use a very simplified image processing system which implements "mini character recognition." Each of the $8 \times 8 = 64$ elements, or *pixels* (from *picture element*), is a threshold automaton connected to all of the others. The state of the pixel is represented by its color, where black represents state 1 and white state −1. In this context, the configurations of the automata are also called "patterns." The three reference patterns represent the letters X, b, and K. The pattern X, for example, corresponds to the configuration shown in figure 5.2:

Figure 5.3 represents some of the results obtained from the evolution of the network constructed by using references X, b, and K to set the synaptic weights according to the Hebb rule. We observe the following:

- The three references defined in figure 5.3.a are invariant.
- We have shown in figure 5.3.b several patterns which belong to the basin of attraction of the reference "X." We can see that some small perturbations in the reference, obtained by inverting a few of its pixels, are "corrected" by the convergent dynamics of the system.
- We can see from figure 5.3.c that there is another invariant pattern which is almost the opposite of the pattern "K." Indeed, the Hebb rule with a threshold of 0 is symmetrical for pattern inversion (the inverse, or the opposite of a configuration is obtained by inverting the signs of all of the states of the automata, or equivalently, by exchanging 1's and 0's in a 0,1 representation). Each time a pattern is memorized, the inverse pattern is also memorized. The slight difference between the observed attractor and the inverse of the pattern "K" is due to the small asymmetry introduced by the way in which we treat

the case where the argument of the transition function is equal to 0. It also illustrates the fact that the attractor is not necessarily exactly the same as the reference.

- Figure 5.3.d illustrates that translating a reference is not a small perturbation. It requires the inversion of a large number of pixels of the reference, and the translated reference is not recognized by the network, as can be seen in the figure.

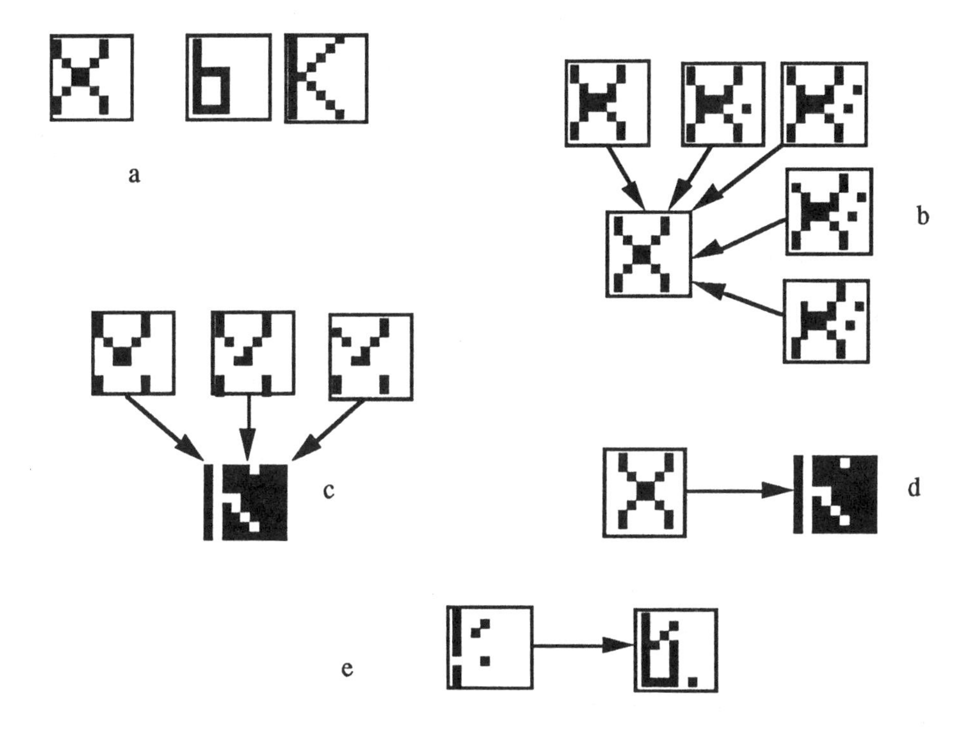

FIGURE 5.3 Examples of the convergence of configurations toward attractors in a Hopfield system. Each square contains 8 × 8 boxes (or pixels), each of which represents the state of an automaton (the black boxes represent automata in state 1, the white boxes represent automata in state -1). The arrows point from the patterns taken as initial configurations toward the corresponding attractors. The three squares in figure 5.3.a representing the letters X, b, and K were chosen as references. They determined the choice of the network connections T_{ij} and are dynamically invariant. We can see that in the iterations shown in figure 5.3.b, some small perturbations in the pattern X were corrected. Other, more significant errors were not corrected (5.3.c). Translating a pattern leads to a different attractor (5.3.d). We also notice the existence of an attractor which is a combination of the two patterns b and K (5.3.e).

- Finally, we notice in figure 5.3.e the appearance of an attractor which is a combination of "b" and "K." The existence of this spurious attractor will be explained in chapter 8 (section 8.4).

5–5 The Signal-to-Noise-Ratio Method

Three methods can be used to predict the behavior of threshold neural networks constructed with the Hebb rule and to explain the results of the simulations. The simplest and most general method, although it is only approximate, is the study of the signal-to-noise ratio, which is explained below. The two other methods, the replica and distance methods, are more rigorous but more complicated, and will be mentioned in chapters 8 and 11.

5–5–1 The Fundamental Rule

The necessary and sufficient condition for a configuration $S(t)$ to be succeeded by the configuration $S(t+1)$, which can either be the same or different from $S(t)$, can be written in the form of N inequalities, one for each automaton i:

$$0 < S_i(t+1) \sum_j T_{ij} S_j(t) \tag{5.6}$$

These inequalities follow directly from equation (5.1), which defines the transition function. If the weighted sum, the argument of the Heaviside function, is positive, $S_i(t+1) = 1$ and their product is positive. Otherwise, the two terms are negative and their product is still positive.

5–5–2 The Invariance of the References

A configuration S^s is therefore *invariant*, in other words it is its own successor, if the following N inequalities (one for each automaton i) hold:

$$0 < S_i^s \sum_j T_{ij} S_j^s \tag{5.7}$$

The two inequalities (5.6) and (5.7) hold regardless of the choice of the T_{ij}.

If the T_{ij} are given by the Hebb rule, inequality (5.7) becomes:

$$0 < N - 1 + \sum_{s' \neq s} \sum_j S_i^s S_j^s S_i^{s'} S_j^{s'} \tag{5.8}$$

where we have replaced T_{ij} by its expression as a function of the references S^s. The first term, $N-1$, called the ***signal*** term, corresponds to the term in the sum over s' for which $s' = s$. Each of the $N-1$ terms of the sum over j is a product of squares of terms equal to 1 or -1, and is therefore equal to 1. It is the large size of the signal term, which is of order N, which guarantees the invariance of the reference.

The double sum includes $q = (N-1)\cdot(m-1)$ terms equal to 1 or -1. If the different sequences are uncorrelated, this sum is 0 on the average. Its standard deviation is the square root of the number of terms, and as long as this quantity is less than $N-1$, this equality is likely to be true. This term is therefore a *noise* term.

More precisely, the probability that the sum of q terms equal to $+1$ or to -1 is less than x is given by the function $Q\left(x/\sqrt{q}\right)$, defined by:

$$Q\left(\frac{x}{\sqrt{q}}\right) = \frac{1}{\sqrt{2\pi q}}\int_{-\infty}^{x} \exp\left(-\frac{x^2}{2q}\right)dx \tag{5.9}$$

The graph of the function Q is shown in figure 5.4. The probability that one of the inequalities (5.8) is true is therefore given by:

$$P = Q\left(\sqrt{\frac{N-1}{m-1}}\right) \tag{5.10}$$

Consequently, by choosing a large argument for the error function, P can be made arbitrarily close to 1. Let y be the argument of the function $Q(y)$ which corresponds to the probability P°. y decreases as m increases. The maximum number of references which yield a probability of invariance greater than P° is therefore:

$$m = 1 + \frac{N-1}{y^2} \tag{5.11}$$

To a first approximation, m varies linearly with N.

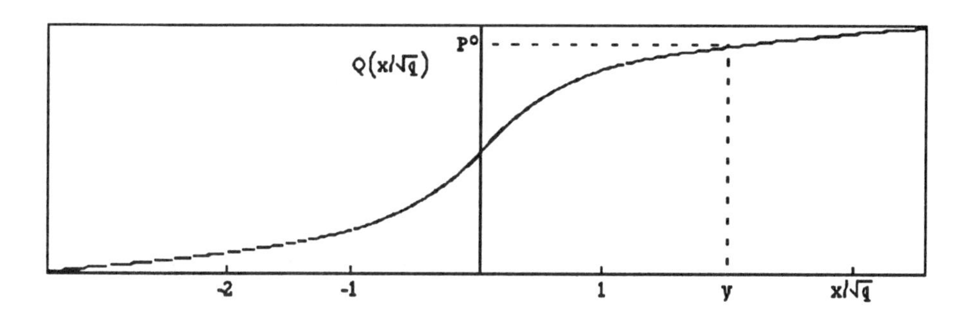

FIGURE 5.4 Probability $Q\left(x/\sqrt{q}\right)$ that the noise term q is less than the signal term x.

"APPROXIMATE" INVARIANCE In fact, there are as many inequalities as there are automata, and condition (5.11) stated above only guarantees the invariance of $P^o N$ of the N references, or that the attractor of the network only differs from the reference in $(1-P^o)N$ automata. In other words, if we accept that the attractor can be a little different from the reference, we can expect to obtain a maximum capacity m which is proportional to the number of automata. This is the first scaling rule concerning the relationship between the number of references and the number of automata. The proportionality coefficient depends on the error with respect to the reference, e.g. the number of automata which have different states in the reference and in the attractor. In practice, when the number of references reaches $0.14N$, even though the error is still only of the order of a hundredth of the number of automata, we observe the "14% catastrophe." For greater fractions, no reference can be stored and the attractors are unrelated to the references. This discontinuity

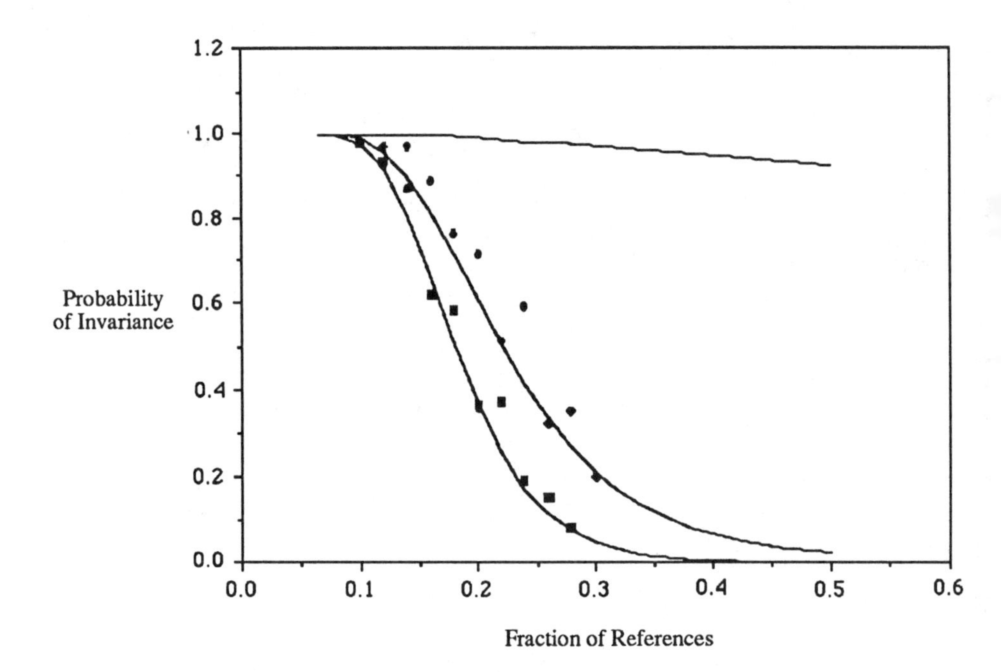

FIGURE 5.5 The probability that a reference configuration is invariant decreases with the ratio of the number of references to the number of automata. The spots and the squares are the results of numerical simulations for 50 and 100 automata, respectively. The solid lines are graphs of the theoretical equation (5.12). The top trace corresponds to the probability of invariance of a single automaton according to equation (5.10).

in the behavior appears in numerical simulations for networks with a large number of automata (at least several hundred). This important result is predicted by the mean field method, which we will discuss in chapter 8 (see sections 8.3.1 and 8.4).

"STRICT" INVARIANCE To guarantee that the N automata of the reference configuration are invariant, N inequalities must be satisfied. If we assume that the double sums are independent, which is a reasonable simplification, we can calculate P_N, the probability that the N inequalities hold:

$$P_N = Q^N\left(\sqrt{\frac{N-1}{m-1}}\right) \tag{5.12}$$

Since P is less than 1, P_N is less than P, and the same probability P^o is obtained for a larger argument of the error function, and thus for a smaller value of m (Fig. 5.5).

5-5-3 Scaling law

If we replace $Q(x)$ by its value in the limit of large x

$$Q(x) \underset{\text{large } x}{\longrightarrow} 1 - \frac{\exp\left(-x^2/2\right)}{\sqrt{2\pi}x} \tag{5.13}$$

we obtain the following relationship between m, N, and P_N:

$$\frac{N-1}{2(m-1)} = \log\left(\frac{N\cdot(m-1)}{\sqrt{N-1}}\right) - \log\left(\sqrt{2\pi}(1-P_N)\right) \tag{5.14}$$

which we can simplify by choosing a value for P_N which nullifies the last term ($P_N = 0.6$). For large values of N, to a first approximation m is given by:

$$m = 1 + \frac{N-1}{2\log N} \tag{5.15}$$

This equation has been confirmed by numerical simulations. This $N/\log N$ scaling law is the condition for strict invariance of the references.

5-5-4 Attraction of the References

One of the most interesting properties of these networks is that configurations which are near a reference converge toward it. By "nearby" configurations we mean configurations which only differ in a few places from the reference. Let $S^{s''}$ be a configuration which only differs in d places from the nearby reference S^s. d is the Hamming distance between both configurations. For an unmodified automaton i, the condition (5.6) for returning in a single iteration toward the reference can be written:

$$0 < N - 1 - 2d + \sum_{s' \neq s''} \sum_{j} S_i^{s'} S_j^{s'} S_i^{s''} S_j^{s''} \tag{5.16}$$

by using the Hebb rule. The signal term has been reduced by $2d$. Although the random sum has been modified, it has the same number of terms. Its probability of being smaller than the signal term is therefore:

$$Q\left(\frac{N-1-2d}{\sqrt{(N-1)(m-1)}}\right) \tag{5.17}$$

A modified automaton i returns in a single iteration to the reference if:

$$0 < N + 1 - 2d + \sum_{s' \neq s''} \sum_{j} S_i^{s'} S_j^{s'} S_i^{s''} S_j^{s''} \tag{5.18}$$

A precise calculation of the probability of returning to the reference would require that we take both inequalities (5.16) and (5.18) into account, as well as the fact that for each update of a modified automaton, the distance to the reference decreases. We will simply settle for a lower bound of the probability, which is:

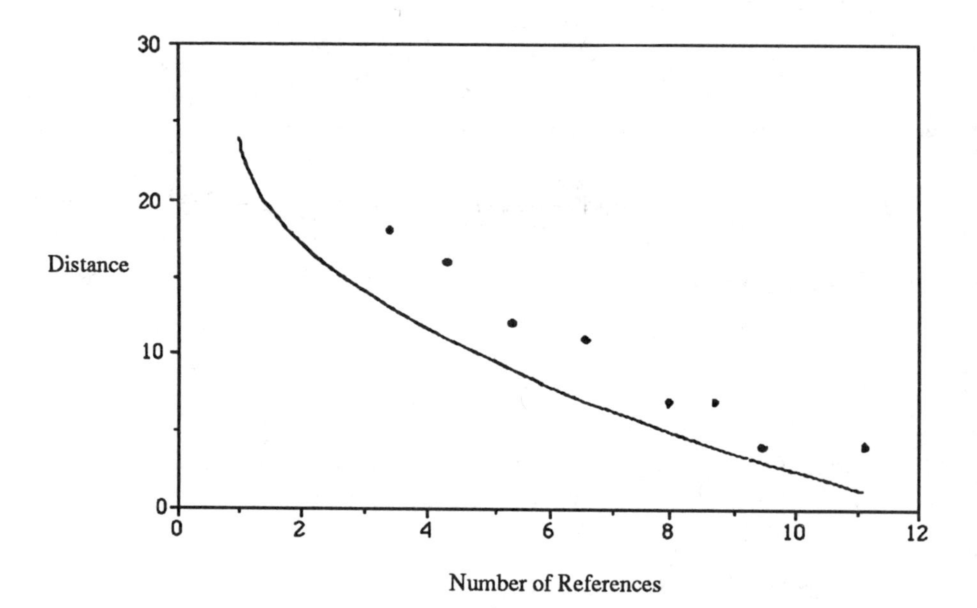

FIGURE 5.6 Distance to the reference which corresponds to a 50% probability of returning to the reference, as a function of the number of references, for a network of 50 automata. The points are the results of numerical simulations, and the solid line is the graph of equation (5.20).

$$Q^N\left(\frac{N-1-2d}{\sqrt{(N-1)(m-1)}}\right) \tag{5.19}$$

A simple way to obtain the relationship between the number of references and the distance corresponding to a given probability of attraction is to equate the argument of the function Q given above with the argument which was used previously for strict invariance with a maximum number of sequences m^* (equation (5.12)). In this manner we obtain the following equation:

$$\left(1-\frac{2d}{N-1}\right)^2 = \frac{m-1}{m^*-1} \tag{5.20}$$

The attraction distance d varies between $(N-1)/2$ for a single reference and 0 when the number of references is at a maximum (Fig. 5.6).

5–6 The Hopfield Network as a Model of a Cognitive System

The Hopfield network is but a very approximate model of the cognitive function of the brain. Its greatest value is that it goes from the cellular level, modeled by the automata, to the cognitive level of the brain, modeled by the collective dynamical properties of the network. Let's point out a few of its properties.

The information is *distributed* in the network. This notion has not always been perceived to be a natural one. A number of cognitive models of the brain have been based on variations of the "grandmother cell" model. In these models, a particular piece of information is assumed to be coded for by one or several cells in the brain. If a person recognizes his grandmother's face among several different faces, it is because a small group of cells which respond selectively to the grandmother have been stimulated. In the Hopfield model, on the other hand, a piece of information is represented by a configuration of neuronal activity, and recognition is interpreted as the fact that this configuration is an attractor of the network.

This method of storing information is that of an *associative memory.* In a classical computer, the memories are said to be random-access, which means that there is no relationship between the address where information is stored and the information itself. One must have the exact address of the complete information in order to access it, and any partial information is useless. In the case of an associative memory, the complete information can be reconstructed from the partial information. This is indeed the case when we give the automata network an initial configuration which is close to the reference: the dynamics of the network restores the reference.

Another property likens this network more to the brain than to a classical computer: its *robustness.* If we eliminate a fraction of the automata in the network, the attractors are scarcely modified. This is true for the same reasons that guarantee the convergence of initial configurations toward the nearest references. This property of progressive deterioration of performance is similar to that of holograms in optics. It also is reminiscent of the fact that, in spite of the loss of thousands of neurons each day, the performance of the brain deteriorates only very slowly with age. Comparable destructions in a von Neumann classical computer would very quickly have disastrous consequences.

Finally, the learning procedure is purely local, and neither it nor the recognition procedure require the intervention of a central processor.

On the other hand, one of the differences between the brain and the Hopfield model is the hypothesis of the homogeneity of the network. Today, no one would claim that the full connectivity used in the model is analogous to the brain, which we know is highly structured. This hypothesis could, however, apply to the cortical columns, which are fully interconnected subsets of cells in the cerebral cortex.

References

J. Hopfield's model is described in "Neural Networks and Physical Systems with Emergent Collective Computational Abilities," *P.N.A.S. USA*, vol. 79 (1982), 2554–2558. This paper is remarkable both in terms of its clarity and the questions which it raises. The problems of going from a biological neuron to its formal representation are discussed in part by P. Peretto in a somewhat difficult paper "Collective Properties of Neural Networks: a Statistical Physics Approach," *Biol. Cybern.*, vol. 50 (1984), 51–62.

SIX

Beyond the Hopfield Model

The Hopfield model is in fact an academic exercise—instructive from a theoretical point of view, but not very useful for practical applications. As a model of the brain, it is based on unrealistic simplifications, such as:

- full and symmetric connectivity, based on the Hebb rule; and
- the choice of uncorrelated, random references.

As a technological recognition and memory device, it is limited by:

- a capacity in the number of references which is a small fraction of the number of automata ($0.14N$ if we tolerate a small error rate of the order of 1%); and
- the dramatic deterioration of performance as soon as we attempt to exceed the maximum capacity.

Our aim in this chapter is to describe variants of the Hopfield model and analogous techniques to overcome some of these limitations. We will start by describing different ways to increase the capacity of the network, and go on to discuss the effects of weakening certain hypotheses on the connectivity which are incompatible with neurobiological observations. Finally, we will study variants which have sequences of references as attractors.

6–1 Increasing the Capacity

The maximum number of bits of information which can be stored in a Hopfield network is given by:

$$mN = 0.14N^2 \tag{6.1}$$

where m is the number of patterns and N is the number of automata.

On the other hand, the set of connections can store on the order of

$$\frac{N^2}{2}[\log_2 A + 1] \tag{6.2}$$

bits of information, where the expression in brackets corresponds to the information, measured in bits, which is contained in a connection with a maximum positive or negative (+1) amplitude A, encoded in binary ($\log_2 A$).

The first quantity is smaller than the second, which indicates that the storage capacity of the network can be increased. Two different approaches can be used to obtain an information content which is closer to the theoretical limit:

- the perceptron algorithm and variations thereon, and
- algebraic methods.

Both methods were in fact developed prior to the Hopfield model.

6–1–1 The Perceptron Algorithm

The perceptron, one of the first automata-based "machines," was introduced for pattern classification, and can learn through the modification of the synaptic weights of threshold neurons. This will be discussed in chapter 7. As its name implies, the algorithm described below was first proposed for this machine, but is more generally useful.

We start with any matrix T_{ij}, such as (but not restricted to) the one used in the Hopfield model. For each automaton i and for each sequence s, we test the invariance condition by evaluating the sign of the sum Σ

$$\Sigma = \sum_j T_{ij} S_i^s S_j^s \tag{6.3}$$

If the sum Σ is positive, invariance is guaranteed (from equation (5.7)). Therefore T_{ij} remains unchanged.

Otherwise, the set of connections to the automaton i from the automata j are modified according to the prescription:

$$T_{ij} \longrightarrow T_{ij} + S_i^s S_j^s \tag{6.4}$$

where the arrow indicates that the term on the left is replaced by the one on the right.

The algorithm is applied to the set of all the automata, for all of the sequences, until all the sums Σ remains positive. It can be shown that the upper limit of the capacity, for uncorrelated random patterns, is $2N$. Notice that the connection matrix is no longer necessarily symmetric.

The same technique allows us to guarantee the attraction of the references. In order for a distant pattern to converge to a given reference in a single step, the sum Σ must be positive and as large as possible, according to the reasoning given in chapter 5, equation (5.16). This ensures that changing the state of a few automata is not sufficient to change the sign of Σ. One way to meet this condition is to verify that the following inequalities are true:

$$\sum_j T_{ij} S_i^s S_j^s - K > 0 \tag{6.5}$$

where K is a constant which guarantees the convergence within a radius proportional to K. [The properties of the network are unchanged if all of the T_{ij} are multiplied by a constant factor; therefore, the radius of convergence depends on the amplitude of K relative to the T_{ij}. Frequently the T_{ij} are normalized so that for each automaton i, the sum over j of their squares is equal to N, which makes the T_{ij} of the order of 1. In this case, the radius of convergence is of the order of $K\sqrt{N}$]. The perceptron algorithm can be used in the same way as it was for invariance, but by testing the inequality (6.5).

One variation, which is the optimal algorithm, yields the largest radii of convergence while modifying the T_{ij} only in the most unfavorable cases. For each automaton i, we find the most unstable reference(s), that is to say the one with the smallest sum Σ, and we apply the update rule (6.4) until all of the inequalities are true, with the largest possible K.

GEOMETRICAL INTERPRETATION We will reason independently about each automaton i, in an N-dimensional space. For each reference s and each automaton i there is a vector ${\eta_i}^s$ with components ${\eta_{ij}}^s$, such that

$$\eta_{ij}^s = S_i^s S_j^s \tag{6.6}$$

If we leave out the index i, the convergence condition (6.5) can be written

$$\sum_j T_j^s \eta_j^s > K \tag{6.7}$$

This implies that in the N-dimensional space of the $\{T_j\}$, the projection of the vector T onto the vector η^s must be positive and as large as possible. In order for this condition to hold for all of the η^s, the set of η^s must be contained in a cone whose apex forms an acute angle. If the vectors η^s are randomly chosen, it can be shown that the cone can contain at most $2N$ vectors. When condition (6.7) is not met, the perceptron algorithm brings the vector T_j inside the cone, until all of the projections are positive and as large as possible (Fig. 6.1).

According to this geometrical interpretation, the Hebb rule consists of using the barycenter of the η vectors as the vector T. This is not the best choice because of the fluctuations in the orientations of the vectors η. It can be shown that the optimal algorithm takes into account the vectors η which are the farthest from the barycenter. The final vector T is then the axis of the cone containing the vectors η (the cone with the smallest angle). On the other hand, by using the algebraic method described in the following section (6.1.2), the resulting vector is the axis of the cone whose boundary contains all of the vectors η.

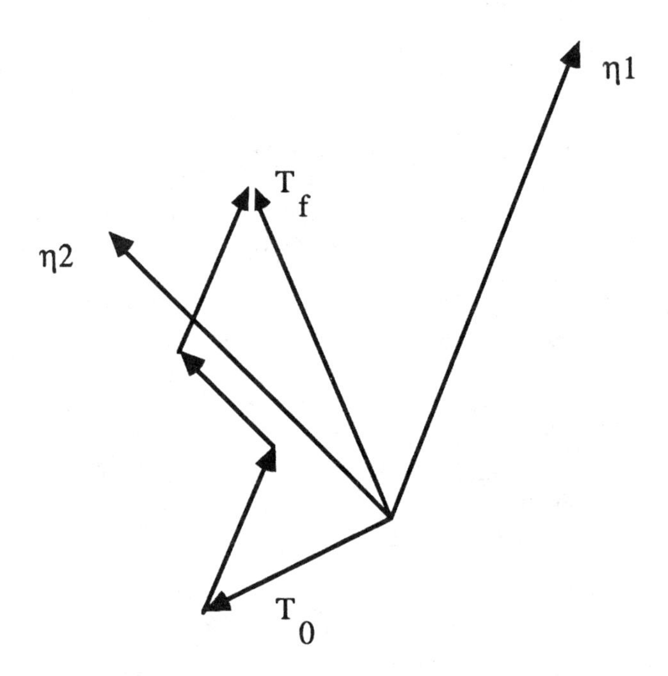

FIGURE 6.1 Geometrical interpretation of the perceptron algorithm. T_0 is a connection vector which is initially ill-oriented. At each iteration, we add to it segments parallel to the references η^1 and η^2 until it "straightens out" into the final vector T_f, which, after learning, has only positive projections onto the references.

6–1–2 Algebraic Methods

Until now we have worked within the framework of automata, or logical arithmetic units. We can, however, address the problem of recognition by networks of analog units with continuous variables. We will summarize this approach, developed by T. Kohonen and others, in order to compare it with the discrete approach.

THE ASSOCIATION PROBLEM Let $\{S\}$ and $\{Y\}$ be two sets of vectors which we would like to pairwise associate. These vectors can be images, or patterns, for example, composed of pixels whose intensities vary continuously between 0 and 1. The (S, Y) pairs of patterns are associated by an unknown matrix $\mathbf{T}$ to be determined, such that

$$Y = \mathbf{T}S\,. \tag{6.8}$$

(Bold-faced terms represent matrices. If Y is of dimension N, equation (6.8) corresponds to N scalar equations.)

If Y is different from S, we speak of hetero-association. If S and Y are identical, we speak of auto-association. The auto-association problem is equivalent to the problem of ensuring the invariance of the patterns S under multiplication by $\mathbf{T}$. At this stage, the problem differs from the Hopfield problem in that:

- the patterns are vectors with real instead of binary components;
- the only operation is multiplication by the matrix $\mathbf{T}$, which is analogous to a matrix of synaptic weights, except that at this point there is no thresholding operation, as there is for thresholding automata, nor are there multiple iterations. The operation is performed only once.

In order to calculate $\mathbf{T}$, we have m equations which relate the input patterns to the output patterns. Grouping these equations we can establish the following matrix relationship:

$$\mathbf{Y} = \mathbf{TS}\,. \tag{6.9}$$

In the $\mathbf{S}$ and $\mathbf{Y}$ matrices, the rows correspond to the analog units and the columns to the patterns. (Equation (6.9) corresponds to $N \cdot m$ scalar equations.) If the number of patterns is equal to the number of analog units, this becomes a classical linear algebra problem which can be solved by inverting the matrix $\mathbf{S}$. If the number of independent patterns is greater than the number of automata, the problem has no solution in general. What follows is a discussion, with no proofs, of the case where the number of patterns is smaller than the number of automata, and consequently the inversion of the matrix $\mathbf{S}$ is not possible.

THE PSEUDO-INVERSE The notion of an inverse matrix, needed to solve the equations (6.9), can be generalized in the case of rectangular matrices to that of a pseudo-inverse. The following three conditions define the pseudo-inverse $\mathbf{S}^+$ of a matrix $\mathbf{S}$:

- $\mathbf{S}\mathbf{S}^+\mathbf{S} = \mathbf{S}$
- $\mathbf{S}^+\mathbf{S}\mathbf{S}^+ = \mathbf{S}^+$
- $\mathbf{S}\mathbf{S}^+$ and $\mathbf{S}^+\mathbf{S}$ are symmetric.

The existence and uniqueness of the pseudo-inverse can be derived. If the columns of the matrix $\mathbf{S}$ are linearly independent, the pseudo-inverse can be written:

$$\mathbf{S}^+ = (\mathbf{S}^t\mathbf{S})^{-1}\mathbf{S}^t \tag{6.10}$$

where $\mathbf{S}^t$ is the transpose of $\mathbf{S}$.

Equation (6.9) can be solved exactly for $\mathbf{T}$ if:

$$\mathbf{Y}\,\mathbf{S}^+\mathbf{S} = \mathbf{Y} \tag{6.11}$$

which is the case for linearly independent patterns, and also for auto-association. The general solution for equation (6.9) can then be written:

$$\mathbf{T} = \mathbf{Y}\mathbf{S}^+ + \mathbf{Z}(\mathbf{I} - \mathbf{S}\mathbf{S}^+) \tag{6.12}$$

where $\mathbf{Z}$ is any matrix. The solution with the smallest quadratic norm is:

$$\mathbf{T} = \mathbf{Y}\mathbf{S}^+ \tag{6.13}$$

(the quadratic norm of a matrix is the sum of the squares of its elements).

If condition (6.11) does not hold, solutions (6.12) and (6.13) are but approximate solutions that minimize the quadratic norm of the error matrix, $\mathbf{Y} - \mathbf{T}\mathbf{S}$.

In the case of auto-association, for linearly independent patterns, we obtain

$$\mathbf{T} = \mathbf{S}(\mathbf{S}^t\mathbf{S})^{-1}\mathbf{S}^t \tag{6.14}$$

and the matrix element T_{ij} of $\mathbf{T}$ can be written:

$$T_{ij} = \frac{1}{N}\sum_{s}\sum_{s'} S_i^s S_j^{s'} Q_{ss'}^{-1} \tag{6.15}$$

where $\mathbf{Q}$ is the matrix of the crossed projections of the patterns, where the matrix elements $Q_{ss'}$ are given by:

$$Q_{ss'} = \frac{1}{N}\sum_{i} S_i^s S_i^{s'} . \tag{6.16}$$

$\mathbf{T}$ can be interpreted as the matrix of projections onto the vector subspace generated by the patterns. The only condition necessary to carry out these calculations is the ability to invert the matrix $\mathbf{Q}$. Therefore, the patterns must be linearly independent. This is a much less restrictive condition than the condition in the Hopfield model that the patterns be uncorrelated, which is in fact an orthogonality condition. It follows that the upper limit of the number of patterns is N, which is clearly an improvement. Note that in the special case in which the patterns are orthogonal, $\mathbf{T}$ is given by the Hebb rule.

ALGORITHMS TO CALCULATE THE CONNECTION MATRIX In the general case, for arbitrary patterns, calculating **T** requires inverting a matrix which can be enormous when we are dealing with a practical application. There also exist exact iterative methods to calculate $\mathbf{S}^+$, such as for example the Gréville method. One can also settle for approximate iterative methods, such as the Widrow-Hoff method.

THE WIDROW-HOFF METHOD The simplest idea is to look for a matrix **W**, *a priori* unknown, to minimize a quadratic error function E. The contribution to the error by each unit of each pattern is the square of the difference between the given pattern Y and the product of S and **W**, the approximation of the matrix **T**. The error function E, which is the sum of the squares of the errors contributed by each of the elements i, for each pattern s, is given by:

$$E = \sum_{s=1}^{m} \sum_{i=1}^{N} \left(\sum_{j=1}^{N} W_{ij} S_j^s - Y_j^s \right)^2 . \tag{6.17}$$

We can then find the minimum of E with respect to the W_{ij} by using a gradient descent method (see the appendix):

$$\mathrm{grad}_w E = 2(\mathbf{W}S - Y)S^t . \tag{6.18}$$

Starting with an arbitrary matrix of synaptic weights **W**, at the kth iteration step we modify **W** according to:

$$\mathbf{W}_k = \mathbf{W}_{k-1} - \lambda_k(\mathbf{W}_{k-1}S - Y)S^t \tag{6.19}$$

where λ_k, the control parameter of the algorithm, decreases with each iteration. Note the similarity to the perceptron algorithm: the synaptic weights are modified only if the calculated output is different from the desired output.

The frequently used Widrow-Hoff method consists of updating the matrix **T** after the presentation of each pattern instead of waiting for all of the patterns to have been presented.

PRACTICAL CONDITIONS FOR RECOGNITION Since multiplication by **T** is a linear operation, the output of a linear superposition of references is also a linear superposition of outputs. Therefore, there is no convergence toward one of the patterns. Convergence can be ensured if nonlinearities are introduced. In practice, we go back to using iterative methods on a network. The calculated vector $\mathbf{T}S$ can be thresholded, for example. Each of its components serves as the input of a threshold automata whose output is 1 if the sum is positive, and -1 otherwise. The application consisting of multiplication by **T** followed by a thresholding operation is then iterated. This method then differs from the Hopfield method only in the calculation of the synaptic weights.

It seems, in fact, more efficient to use analog units which compress the linear combination of the inputs $A = \sum T_{ij} S_j$ in the $[-1, +1]$ interval rather than threshold units. Two possible functions are shown in figure 6.2. The function represented on the left is piecewise linear and is given by:

$$
\begin{aligned}
f(A) &= -1, && \text{if } A < -1 \\
f(A) &= A, && \text{if } -1 < A < 1 \\
f(A) &= 1, && \text{if } A > 1
\end{aligned}
$$

The graph on the right represents the hyperbolic tangent function:

$$f(A) = \tanh(A) = \frac{e^{A} - e^{-A}}{e^{A} + e^{-A}} \tag{6.20}$$

For networks of analog units, as for networks composed of logic units, the result obtained by one application is iterated until the attractor has been reached.

The methods based on the algebraic approach are more powerful and better adapted to the concrete problems of recognition and associative memory than the Hopfield model. These applications, as well as many others, are described in T. Kohonen's book. Numerous variations have been developed. The substantial increase in the number of attractors comes with some disadvantages:

- If the number of patterns approaches the number of automata, the basins of attraction are reduced practically to the references.
- The exact algorithms used to calculate $\mathbf{T}$ are not local. In order to increment a connection, the Hebb algorithm only needs to know the activities of the two connected cells. The pseudo-inverse requires a knowledge of the projection matrix, which involves the activity of the entire network. This non-locality requires information transfer through the network and therefore represents a disadvantage for practical implementations or for a model of the brain. Only the Widrow-Hoff algorithm preserves the notion of locality.

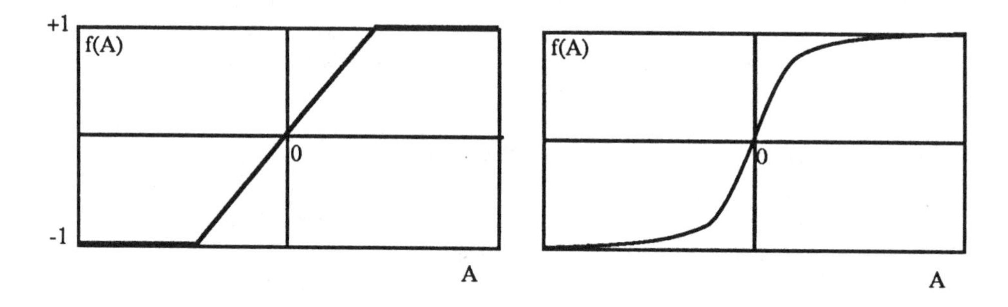

FIGURE 6.2 Compression functions. These functions compress the interval of variation of the weighted sum of the inputs A into the interval $[-1, +1]$.

6–1–3 Changing the Connectivity

Can the memory capacity be increased even further? The answer is yes, but only by changing certain construction principles.

NONLINEAR CONNECTIONS The capacity of the network can be increased by taking as input signals multilinear combinations of the states of the other automata. We will examine the case of bilinear signals using the notation and the signal-to-noise-ratio approach developed in chapter 5.

$x_i(t)$, the state of automaton i at time t, is given by:

$$x_i(t) = Y\left[\sum_k \sum_j T_{ijk} S_j(t-1) S_k(t-1)\right] . \tag{6.21}$$

The ternary connections are calculated from a generalization of the Hebb rule:

$$T_{ijk} = \sum_s S_i^s S_j^s S_k^s . \tag{6.22}$$

The analysis of the signal-to-noise ratio is therefore easily applicable. The invariance condition of the references can be written:

$$S_i^s \sum_j \sum_k T_{ijk} S_j^s S_k^s > 0 \tag{6.23}$$

which can be decomposed into:

$$N^2 + \sum_{s'} \sum_j \sum_k S_i^s S_j^s S_k^s S_i^{s'} S_j^{s'} S_k^{s'} > 0 . \tag{6.24}$$

The first signal term is greater than the standard deviation noise $N\sqrt{m}$ only if

$$m < \alpha N^2 .$$

(It can even be shown by other methods that α is of the order of 0.08.) Therefore the capacity of the network is proportional to the square of the number of automata, as long as a small but finite error percentage is tolerated. Of course, nothing is free! The total number of connections has been considerably increased: it is now of the order of N^3. In general, the number of attractors of a threshold automata network constructed using a "Hebb-like" learning rule varies as the connectivity of the automata. We will see other examples of this later.

MODULAR ARCHITECTURE Thus far we have only discussed homogeneous networks. By structuring a network into independent subnetworks, it is possible to obtain an exponential number of attractors.

Assume N automata distributed into N/p subnetworks of p automata each. Each subnetwork has of the order of αp independent attractors, and the number of different attractors of the global network is:

$$(\alpha p)^{N/p} .$$

Of course, the division process can be carried out even further and the subnetwork can be divided into subsubnetworks.... In the limiting case, a set of N disconnected automata has 2^N stable configurations! But these configurations are not attractive. The subdivision process must stop at a level such that the number of automata of the smallest module guarantees the attractiveness of the reference configurations, and that the multiplicity of the subnetworks allows the greatest possible number of references to be stored.

This modular approach is the one which is observed in the brain, which is subdivided into cortical areas, each of which is further subdivided into columns and microcolumns. From this point of view, the fully connected Hopfield model is more similar to a microcolumn with several tens of thousands of neurons than to the entire brain.

Structured networks have scarcely been studied thus far, either at the application level or at the level of basic principles. One of the problems which emerges is that of the specialization of the subnetworks. The improvement in performance of the system relies on the fact that the attractors of the different subnetworks are different. This difference can be fixed at the outset. In the case of an image processing system, the cells of each subnetwork can be chosen to be sensitive to different characteristic features: the positions of points, the orientations of contours, movements, colors, etc. Consequently, the same image is processed by each subnetwork using different criteria and, after learning, the attractors of the subnetworks are quasi-independent. Again, this seems to be the case for visual area 17 in the cortex, where several maps of the visual field sensitive to different features can be found.

A first approach is to set the structure and the specificity of the subnetworks before any learning, using *a priori* knowledge about the problem to be solved. One can also start with identical subnetworks, or subnetworks which differ only in that the initial conditions are randomly chosen. To prevent the same input patterns from leading to the same attractors, the units of the different subnetworks are connected by fixed inhibitory (i.e., negative) connections, which prevent the cells from being in the same state. In summary, when faced with an organized biological structure such as the brain, one can always ask how much of the organization is acquired and how much is predetermined. Certain experiments, such as those done on cats by D. Hubel and T. Wiesel on the recognition of oriented lines, and certain simulations, such as R. Linsker's work on the structuring of cortex, would suggest that the role of learning is more important than one might have thought *a priori*.

6–2 Avoiding Catastrophe: The Palimpsests

Consider a Hopfield network to which we keep adding patterns. As soon as their number is greater than $0.14N$, the system "loses its memory." None of the previous patterns can now be recognized. In addition, the new patterns presented after the 14% catastrophe cannot be recognized either. Such a fragile system cannot be used directly as a model of learning in the human brain. The first modification of the model which comes to mind is to assume that the synaptic weights are not uniformly malleable. We will therefore assume that at each new presentation p, the variable $\varepsilon(p)$ represents a learning intensity. This can be thought of as a model of attention. The Hebb algorithm can be generalized as follows:

$$T_{ij}(p) = T_{ij}(p-1) + \varepsilon(p) S_i^p S_j^p \,. \tag{6.25}$$

The stability condition of the pth pattern:

$$\sum_j T_{ij}(p) S_i^p S_j^p > 0 \tag{6.26}$$

can be rewritten

$$\sum_j T_{ij}(p-1) S_i^p S_j^p + \varepsilon(p) \cdot (N-1) > 0 \,. \tag{6.27}$$

The second term is a signal term, and the first is a noise term, given that the pattern p is independent of the previous patterns. The stability of the last memorized pattern can therefore always be guaranteed if $\varepsilon(p)$ is chosen to be large enough to make the signal larger than the noise. It is fairly easy to calculate the standard deviation of the noise term and deduce from that the choice of $\varepsilon(p)$. At each step, $\varepsilon(p)$ increased exponentially.

This guarantees the stability of the last sequences learned, but causes the first sequences to be forgotten, since the connection matrix is only weakly dependant on their correlation function. This model can be solved completely, and at each step we can reduce the intensity of the T_{ij} in order to normalize their amplitude.

It is noteworthy that another model which is intuitively simpler and more reasonable in biological terms leads to the same behavior. Let's make the simple assumption that the amplitude of the synapses has an upper bound and that the synapses are either positive or negative. In other words:

$$0 < T_{ij} < A \quad \text{or}$$
$$-A < T_{ij} < 0 \,.$$

Learning is carried out at a constant intensity, but within the limits imposed above. At each presentation, we calculate:

$$T_{ij}(p) = T_{ij}(p-1) + \varepsilon S_i^p S_j^p \,. \tag{6.28}$$

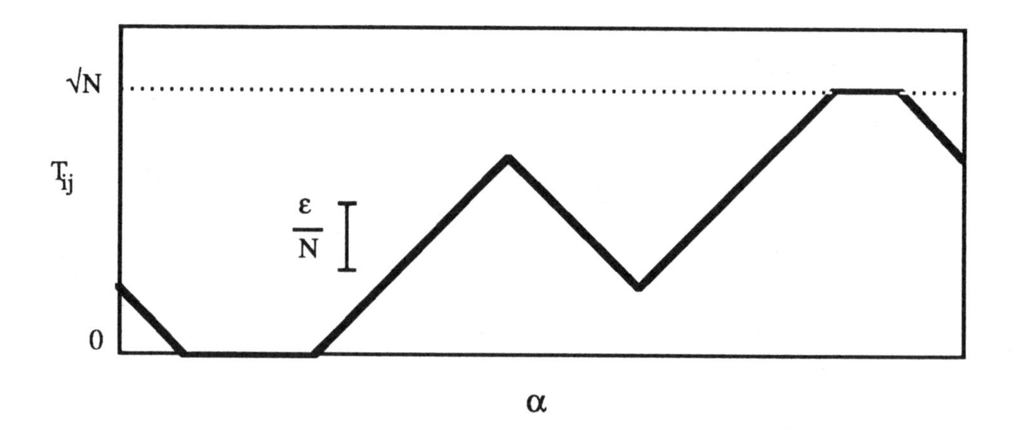

FIGURE 6.3 Evolution of an excitatory connection between 0 and the saturation value $\sqrt{N}$ during learning. (From J.P. Nadal, Ph.D. thesis, Paris, 1987.)

If the new T_{ij} is within those limits, it is conserved. Otherwise, it is given the value of the exceeded limit. Figure 6.3 shows how T_{ij} varies during a learning session.

The following results were predicted theoretically, and verified by simulations. A, the maximum amplitude of the connections, was chosen to be $\sqrt{N}$ in order to make ε independent of N. If ε is too small compared to $\sqrt{N}$, the 14% catastrophe occurs before the connections reach the limit A. On the other hand, if ε is too large, only the last pattern presented is retained. It can be shown that ε_c, the best choice for ε, which is of order 3, allows the αN *last* patterns to be retained with a maximum relative error of about 0.97 ($\alpha = 0.015$). The previous patterns are forgotten. The behavior of this system can be likened to the recovery of ancient parchments by scribes of the Middle Ages, who erased roman manuscripts in order to write over them. The name of the algorithm comes from these recycled parchments, known as palimpsests.

Figure 6.4 shows the evolution of α, the relative capacity of the network in terms of the number of patterns which are retained during the course of the presentation of the patterns. g is the ratio of the number of patterns presented to the number of automata. If ε is too small, the capacity increases at first until it reaches saturation, which is 0.14, and then it decreases to 0. If ε is greater than the critical value ε_c, α does not reach 0.14 on the way up, but after it peaks it will converge to an asymptotic value as g is increased. Figure 6.5 shows the dependance of the asymptotic capacity on ε.

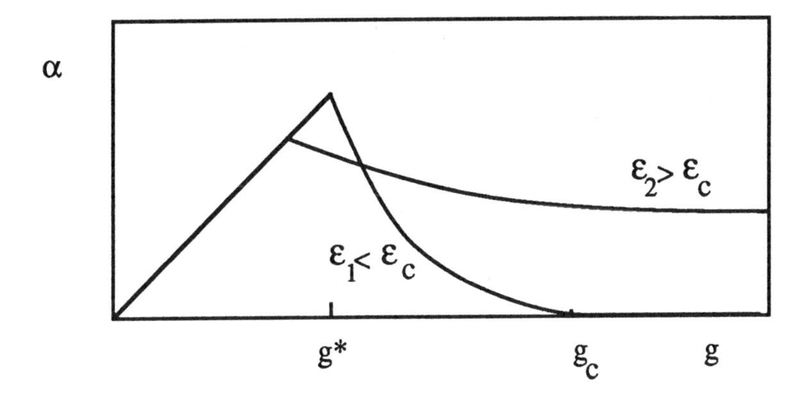

FIGURE 6.4 Capacity of the network as a function of the number of patterns presented for two values of the learning intensity ε_1 and ε_2. (From J.P. Nadal, Ph.D. Thesis, Paris, 1987.)

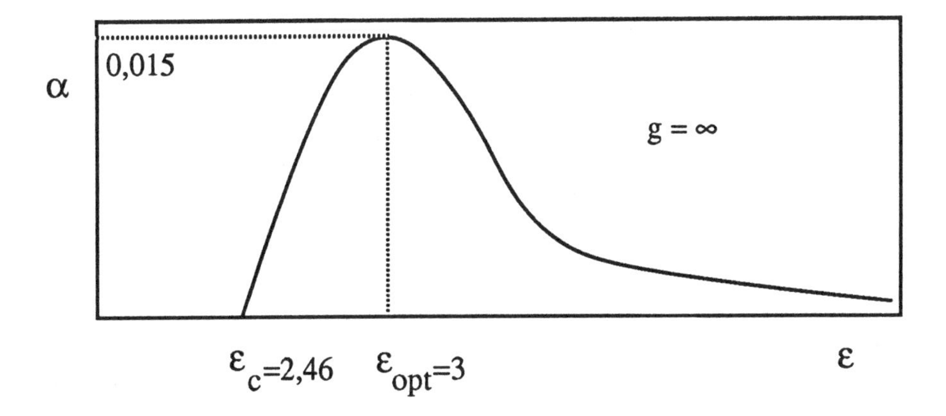

FIGURE 6.5 Asymptotic capacity of the network as a function of the learning intensity. (From J.P. Nadal, Ph.D. Thesis, Paris, 1987.)

The behavior of recalling recent patterns and forgetting old patterns is a reasonable model of short-term memory. In experimental psychology, series of words, simple figures, or numbers are presented to a subject in order to determine the number of objects which can be memorized. Oddly enough, this number is on the average seven with a variation of plus or minus two at most, regardless of the individual and the nature of the objects. This small number can be compared to the small fraction of patterns memorized in the model. It is of interest to note that the model gives us at least a partial answer to the following question: Do we forget because the connection strengths decrease quasi-automatically, or because of distortions brought about by learning new objects? Of course, the model tends to confirm the second hypothesis, which is borne out by recent experiments.

We can also ask questions pertaining to long-term memory. Note that with the same model that we have just described, if we force the connection T_{ij} to be stuck at the first limit it attains, we then favor the memorization of the first patterns.

6–3 Asymmetrical Networks and Dilution

The constraint of symmetrical connections imposed by the Hebb rule is difficult to imagine in a natural system such as the brain. Recall that the perceptron algorithm leads to an asymmetrical connection matrix. We can also start from the Hebb rule and "dilute" the connections by randomly eliminating some of them. This dilution is asymmetrical, since cutting a T_{ij} connection does not imply that the symmetrical connection T_{ji} is also cut. The capacities of the network in terms of the maximum number of attractors and the accuracy of restoration steadily decrease as we increase the percentage of connections which are cut. With 50% of the connections, the maximum number of attractors is $0.09N$ with an accuracy of better than 0.9. A network can even operate in the limit of high dilutions, where the connectivity k (the number of inputs of an automaton) is less than $\log N$. By using a technique developed in chapter 10 known as the distance technique, we can even calculate the performance of these diluted networks theoretically. For a given connectivity k, the maximum number of memorizable patterns is:

$$m - 1 = 0.64k\,. \tag{6.29}$$

q, the projection of the attractor S^a onto the reference S^s is defined to be:

$$q = \frac{1}{N}\sum_i S_i^a S_i^s\,. \tag{6.30}$$

If $q = 1$, the attractor is the same as the reference, and if $q = 0$, they are orthogonal. The difference between q and 1 is therefore an indication of the relative distance between the attractor and the reference. For the very diluted model, q, the projection onto the reference, is given by:

$$q = 1 - \sqrt{\frac{2\alpha}{\pi}}\exp\left(-\frac{1}{2\alpha}\right) \tag{6.31}$$

where $\alpha = (m-1)/k$.

Again, we obtain a maximum capacity which is proportional to the connectivity of the automata. The system does not have a unique fixed point in the neighborhood of the reference. It can be shown that if an initial configuration is near the reference, its average projection onto the reference is given by the previous expression. On the other hand, two initial configurations which are near the reference do not tend to converge toward each other. It can be shown that their average projection tends toward a value which is close to, but not equal to 1. Therefore around the reference there are either several fixed points or a limit cycle.

6–4 Sequences

We will set out to construct a network capable of generating temporal sequences of patterns—in other words, where the interesting attractors are limit cycles rather than fixed points. Nature provides us with many examples of such behavior, from the songs of birds to the sequences in which different muscles are contracted to generate a movement. The difficulty which we have in trying to sing a melody backwards is a good indication of the fact that we memorize sequences rather than a set of states which we could access in any order.

In the case of parallel iteration, the T_{ij} can be chosen as follows:

$$T_{ij}^{t} = \sum_{s} S_i^{s+1} S_j^{s} \,. \tag{6.32}$$

The superscript t of the connection matrix indicates that this is a transition matrix from the s configuration to the $s+1$ configuration.

All of the analysis which we have done previously, such as the signal-to-noise ratio, or the efforts to improve the performance by the perceptron or Widrow-Hoff types of techniques, can be applied to this model provided the correlation functions of the form $S_i{}^s S_j{}^s$ are replaced by the transition functions $S_i{}^{s+1} S_j{}^s$.

The case of sequential iteration is more difficult, but more relevant for biological models. The solution proposed by P. Peretto and J. J. Niez is to use the sum of two terms as the argument of the Heaviside function:

$$\sum_{i} T_{ij}^{s} S_j(t) + \sum_{i} T_{ij}^{t} S_j(t-\tau) \tag{6.33}$$

with

$$T_{ij}^{s} = \sum_{s} S_i^{s} S_j^{s} \tag{6.34}$$

and

$$T_{ij}^{t} = \lambda \sum_{s} S_i^{s+1} S_j^{s} \,. \tag{6.35}$$

The first term guarantees the stability of the sequences, and the second the transition from one sequence to the next. Again, the signal/noise analysis shows that for a certain time τ after a transition, only the first term contains a signal term which ensures stability. After time τ, the second sum also contains a signal term which causes the transition toward the following configuration if λ is greater than 1.

6–5 Temporary Conclusion

In recent years numerous variations of Hopfield systems have been developed, and most of the restrictive hypotheses which gave the model its formal simplicity can be extended without the loss of any of the properties of the model. Of course, this conceptual framework does have certain limitations, some of which can be bypassed by other approaches which we will describe in the following chapters.

One of the most serious limitations is that the system does not allow invariant recognition. The Hopfield model only accepts pointwise distortions of the patterns. Take the example of an image. If a fraction, potentially quite large, of the pixels of the image is modified, the image can still be recognized. On the other hand, distortions of the image such as translation, rotation, dilation, etc. modify a very large number of pixels, and lead to very large Hamming distances to the reference. The distorted image is therefore not recognized. Of course, natural systems recognize distorted patterns, and these properties of invariant recognition are widely sought after in all applications of signal processing, for speech, image recognition, etc. The Hopfield system is essentially only applicable for patterns of characteristic features, rather than images. To use these systems in image processing, one would have to perform some preprocessing on the image, such as centering the image in the case of translations. Another possible form of preprocessing is to extract characteristic features (see section 9.2). In this case, the automata are specialized for specific features, and encode the presence or absence of these features in the image.

References

The performance of the optimal algorithm based on the perceptron algorithm is discussed in a paper by W. Krauth and M. Mézard: "Learning Algorithms with Optimal Stability in Neural Networks," *J. Phys. A: Math Gen.*, vol. 20 (1987): L745.

The algebraic methods are clearly described with a wealth of applications in T. Kohonen's book: *Self-Organization and Associative Memory, Springer Series in Information Sciences*, vol. 8 (Springer Verlag, 1988). The chapter "Automata Networks and Artificial Intelligence" by F. Fogelman-Soulié, P. Gallinari, Y. Le Cun, and S. Thiria, in the collective book *Automata Networks in Computer Science*, edited by F. Fogelman-Soulié, Y. Robert and M. Tchuente (Manchester University Press, 1987), discusses linear and nonlinear formalisms with some comparative evaluations of the different algorithms.

The problem of the differentiation of layers is addressed in a paper by R. Linsker, "From Basic Network Principles to Neural Architecture: Emergence of Orientation Columns," *Proc. Nat. Acad. Sci. U.S.A.*, vol. 83 (1986): 8779–8783.

For the palimpsests, we were inspired by the paper by J. Nadal, G. Toulouse, J. Changeux and S. Dehaene, "Networks of Formal Neurones and Memory Palimpsests," *Europhysics Letters*, vol. 1 (1986): 535–542.

The results on diluted networks are taken from B. Derrida, E. Gardner, and A. Zippelius, "An Exactly Soluble Asymmetric Neural Network Model," *Europhysics Letters*, vol. 4 (1987): 167.

SEVEN
Hidden Units

Classifications

A Hopfield network classifies different categories of inputs according to their basin of attraction: two classes are considered equivalent if they have the same attractor. We have seen that this classification depends on the choice of a measure of closeness in the sense of the Hamming distance. In fact, although the choice of this measure may be of interest in some cases, it is not necessarily relevant in others. In the field of pattern recognition, for example, one is chiefly interested in invariants: two shapes are considered equivalent if they only differ by a translation, although their Hamming distance can be very large.

Our aim in this chapter is to specify the nature of the classifications which can be carried out and learned by Hopfield networks and equivalent networks. We will then propose other types of networks which can solve problems which cannot be solved by Hopfield networks.

7–1 The Perceptron

In fact, the ability of Hopfield networks to discriminate between patterns of input is equivalent to that of the perceptron, a machine which is already over twenty years old. In its simplest version, the perceptron is a network composed of three layers:

- The first layer, or retina, is the input unit layer which receives input patterns to be classified.

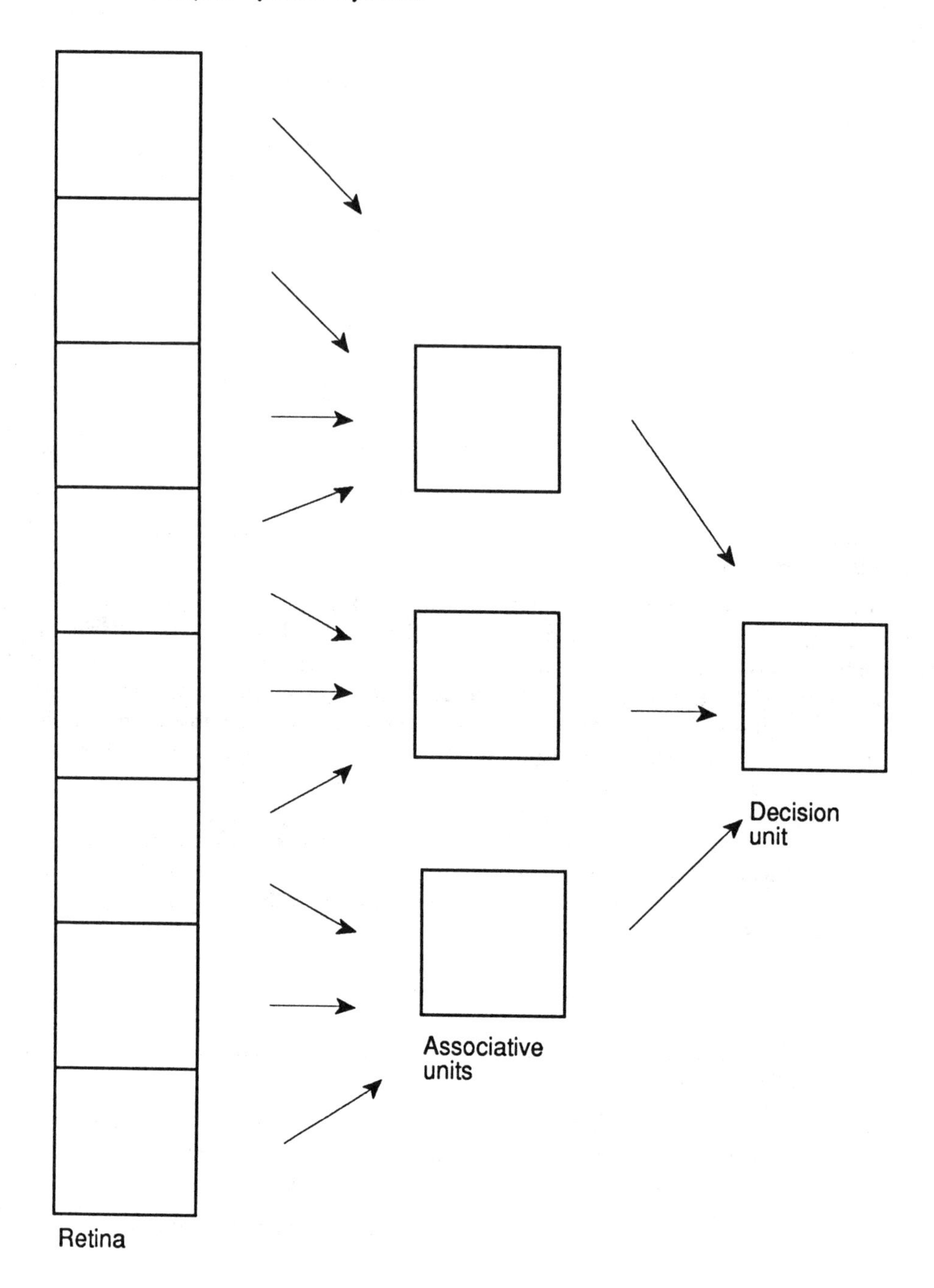

FIGURE 7.1 Diagram of a perceptron.

- The second layer is composed of associative units, such as boolean automata, which receive inputs from the units of the retina. Each unit computes a distinct, fixed boolean function which cannot be modified by learning.
- The last layer is composed of one or more decision units, which are threshold automata with weights that are modified during the learning process (Fig. 7.1).

There is no feedback within this network: information is therefore processed in a single iteration through the successive layers. There is no dynamical iteration toward an attractor as there was in the examples previously described.

Linear Classifiers

Let us first consider the case of a single decision unit. The function Φ computed by this unit is given by

$$\Phi(s) = Y\left(\sum_j T_j f_j(s) - \Theta\right) \tag{7.1}$$

where T_j is the synaptic weight and f_j is the boolean function of the jth associative unit, s is the input pattern, Y is the Heaviside function, and Θ is the threshold of the unit. We can therefore define two classes of input patterns, one being the class of patterns such that Φ is equal to 1.

The first point worth mentioning is that the perceptron is a *linear classifier*. Indeed, the space of the $f_i(s)$ is divided in two regions by the hyperplane defined by the T_j's and the threshold. A single decision unit can only separate two classes on either side of a hyperplane. The separation can of course be refined by adding more decision units. The different classes of inputs are the polytopes separated by the hyperplanes in the space of inputs.

Hopfield networks are based on essentially the same stability criteria as perceptrons. The computation rule for the perceptron:

$$\left(\sum_j T_j f_j(s) - \Theta\right) \cdot \Phi(s) > 0 \tag{7.2}$$

is basically the stability equation for the Hopfield network given in chapter 5. Thus, Hopfield networks are also linear classifiers, and as such inherit the same limitations. Before we go on to discuss the limitations of linear classifiers, let us recall the following interesting result which was derived for the perceptron.

The Convergence Theorem

In chapter 6 we discussed the learning rule for the perceptron. Let us rewrite it here in the notation we defined for the perceptron. We will use the $-1, +1$ representation for the decision unit. The observed output signal is therefore $S = 2\Phi - 1$. Let y be the desired output of the system in response to a given input pattern. Initially, the synaptic weights are random. For each input pattern, the synaptic weights are modified if and only if y and S differ, according to:

$$T_j \longrightarrow T_j + (y - S) f_j(\mathbf{s}) . \tag{7.3}$$

The threshold is also modified by the same algorithm, since it can be treated as an extra associative unit which has no input, and the state of which is fixed at the value -1. Θ is simply the synaptic weight of this unit. The algorithm is applied until all desired and observed inputs coincide.

The perceptron convergence theorem states that if there exists a function which can achieve the chosen classification, then the algorithm will converge in a finite time toward a satisfactory function Φ.

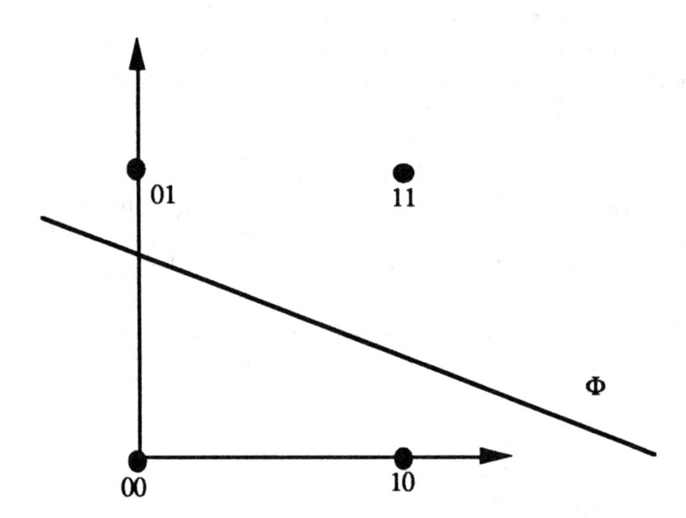

FIGURE 7.2 The thresholding function Φ classifies the four inputs 00, 01, 10, and 11 according to their position relative to the line Φ. Incidentally, we can see on this figure that by rotating the line Φ, thresholding functions can represent 14 of the 16 boolean functions with two inputs. The only exceptions are the XOR function and the EQUivalence function, that is to say the functions with codes 6 and 9.

A Simple Counter-Example

At this stage, it becomes interesting to specify what linear classifiers cannot do. Problems emerge when we try to regroup classes of inputs that were separated by hyperplanes. Consider the case in which we would like to implement the exclusive OR (XOR) function with two input units. In two dimensions, the input space is a plane, and the linear separator is the line given by:

$$T_1x + T_2y = \Theta \tag{7.4}$$

where x is the state of unit 1 and y is the state of unit 2.

This geometrical representation illustrates that it is impossible to implement the XOR function, which would put the 01 and 10 inputs in the same class (Fig. 7.2). This is bad news, since this function detects the presence of a single 1 in the input pattern, irrespective of whether the 1 occurs on the first or second unit, which makes XOR an important function as a model of invariant recognition.

Alternatively, the associative units could be given the task of implementing functions of the XOR type. However, these units are not able to learn. In other words, in making this choice, we would give up the idea of learning by examples.

However, since the work of McCulloch and Pitts, we know that a network of threshold automata is capable of implementing the XOR function. An example of such a network is shown in figure 7.3.

Hidden Units

A difficulty arises when we consider the central unit, or *hidden unit* in figure 7.3. The statement of the problem only specifies the states of the two input units and of the output unit. The state of the middle unit is not specified *a priori*. An algorithm which can vary the synaptic weight of this unit must be able to solve the "credit assignment problem." The activity of a hidden unit can increase or decrease the error between the desired output pattern and the output pattern computed by the network; consequently, how can one vary the synaptic weight of a unit without knowing what its final state will be in the recognition phase of the "educated" network? Recall that all of the algorithms which have been discussed up to this point have assumed that all of the states of the automata in the network are known.

At this stage in our discussion the situation is the following: it would appear that by using hidden units, we could expand the class of problems we can solve, but we still need a new learning algorithm.

Some History: The Perceptron Debate

Before discussing a simple solution (simple now that we know it!) to the previous question, it is of interest to note that this problem, formally posed by M. Minsky and

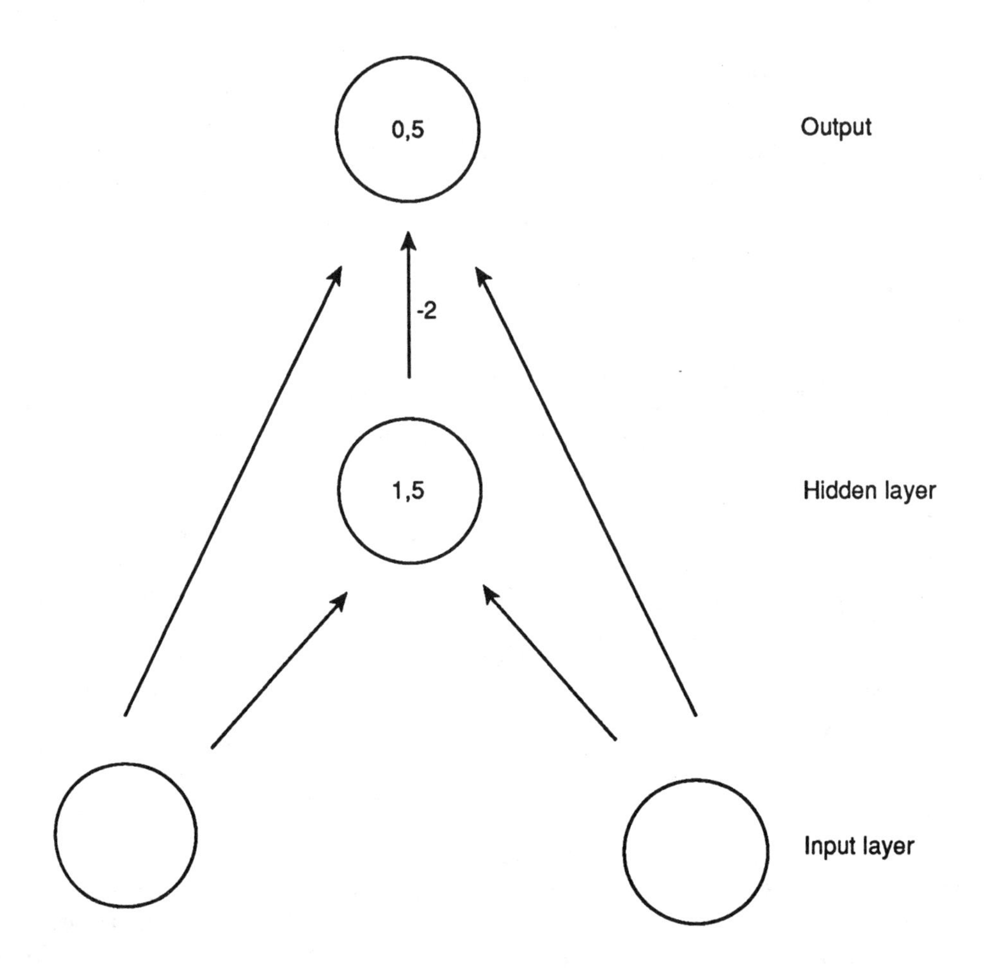

FIGURE 7.3 The XOR function can be implemented with two threshold automata. All of the arrows represent connections with synaptic weights of 1 except the middle arrow, which has a synaptic weight of -2. The numbers inside the circles refer to the thresholds.

S. Papert in 1969, almost completely blocked the development of applied research on networks of automata for more than ten years. To summarize their reasoning:

- Perceptrons are linear separators.
- Functions of the XOR type are necessary for pattern recognition. (One of the examples which was cited as requiring the XOR function was the problem of recognizing the connectedness of a geometrical figure.)
- Networks of perceptrons with only one decision layer cannot learn this type of function.

- We do not know any learning algorithm for layered networks. We will never find any; it follows that we should stop doing research on this subject with no future.

Perhaps perceptrons did not deserve the wild expectations that had been triggered by the initial research of F. Rosenblatt (1962), but neither did they deserve the attacks inspired by the book of M. Minsky and S. Papert (1969).

One should keep in mind that the perceptron controversy was at the center of a debate between analog vs digital machines on the one hand (the perceptron was implemented with variable resistors operated by big motors during the learning phase), and dedicated machines (such as the perceptron) vs general purpose machines on the other hand. Technology and need guided the choice in the early fifties toward digital, general purpose machines. In fact, it was thanks to the simulation possibilities offered by general purpose digital machines that the field has once again been developing since the early eighties.

7–2 Layered Networks

7–2–1 Structure

Today, these networks represent the answer which is best adapted to solving the problems posed by the limits of the perceptron and the Hopfield models: they allow classifications which go beyond linear classifications, and we have access to learning algorithms which enable them to learn and to generalize from examples.

The idea is the following: these are systems in which the signal propagates, in each time interval, from the first, or input layer through successive intermediate layers toward an output layer.

Each unit (with the exception of the input units) evaluates a linear combination A of its inputs from the preceding layer, and compresses this algebraic sum in the interval $[-1, +1]$ (see the examples of compression functions in 6.1.2). The algebraic sum contains a threshold term $-\Theta$. Before the learning phase, each unit is connected to all of the units in the preceding layer. Contrary to classical models of networks of automata, the units are not digital, but analog. The output signal of a unit i is written:

$$S_i = f\left(\sum_j T_{ij} S_j - \Theta_i\right) \qquad (7.5)$$

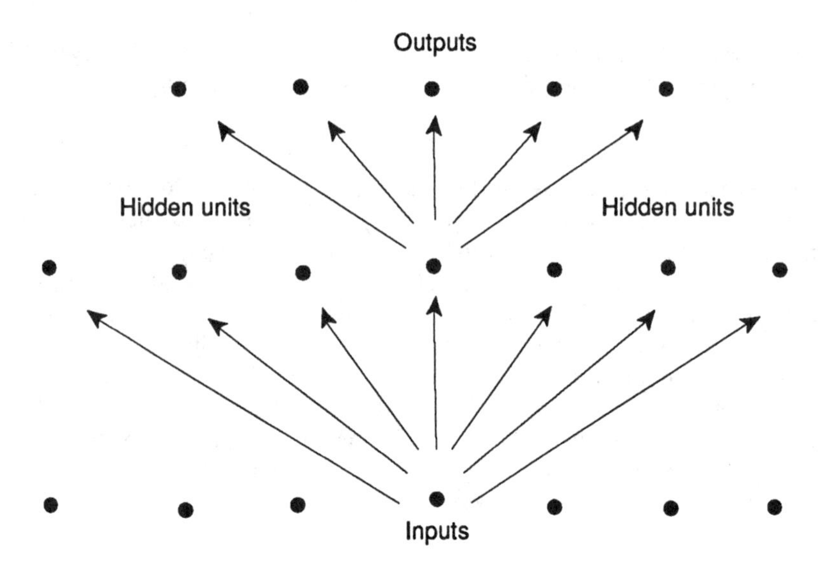

FIGURE 7.4 A layered network.

where f is a compressing function like the ones described in section 6.1.2, such as the hyperbolic tangent function:

$$f(A) = \tanh(A) = \frac{e^A - e^{-A}}{e^A + e^{-A}} \tag{7.6}$$

whose derivative with respect to A can be written in the form:

$$f'(A) = 1 - f^2(A)\,. \tag{7.7}$$

The architecture of this network can vary considerably within this framework depending on the application, since neither the number of units in each layer nor the number of hidden layers is fixed.

Finally, and most importantly, the synaptic weights (and the thresholds) are learned as a function of the patterns presented to the input units and the resulting outputs of the output units. We start with small, randomly chosen weights and apply the back-propagation algorithm, described in the following paragraph (7.2.2). We could of course imagine more general connectivity schemes for this network, but the back-propagation algorithm is applicable only for the framework which we have just described.

7–2–2 Back-Propagation

We want to find the synaptic weights which minimize an error function C:

$$C = \sum_s \sum_i (S_i^s - Y_i^s)^2 \tag{7.8}$$

where Y is the vector of desired outputs and S is the vector of outputs as computed by the network. The i's index the units of the output layer, and the s's index the input pattern.

The method which we will use is a gradient descent method: first we calculate the components of the gradient of the error function C in the space of synaptic weights, and then at each step, each synaptic weight is reduced by an amount which is proportional to the corresponding partial derivative of the error function. As in the case of the perceptron, the thresholds are treated as the synaptic weights between a unit j with a fixed output -1 and the unit i which is being considered.

We begin by choosing a Widrow-Hoff type of iteration: instead of evaluating the gradients for the error function which is the sum of the errors obtained for all of the input patterns, we modify the synaptic weights each time a pattern is presented, taking into account only the error on that particular pattern.

In order to simplify the notation of the algorithm, we will introduce the intermediate variables A_i which represent the weighted sums of the outputs of the units j:

$$A_i = \sum_j T_{ij} S_j \,. \tag{7.9}$$

In general, the partial derivative of C with respect to a synaptic weight involves a product of two terms:

$$\frac{\partial C}{\partial T_{ij}} = \frac{\partial C}{\partial A_i}\frac{\partial A_i}{\partial T_{ij}} = \frac{\partial C}{\partial A_i} S_j \,. \tag{7.10}$$

Consequently, we are interested in determining the derivatives with respect to the intermediate variables A_i, which can then simply be multiplied by the S_j to obtain the derivatives with respect to the corresponding synaptic weights.

THE CHAIN OF DERIVATIVES The calculation of the state of each output unit involves the states of the intermediate units through nested expressions of the same form:

$$S_i = f(A_i) = f\left(\sum_j T_{ij} S_j\right) \tag{7.11}$$

$$S_i = f(A_i) = f\left(\sum_j T_{ij} f(A_j)\right) \tag{7.12}$$

$$S_i = f(A_i) = f\left(\sum_j T_{ij} f\left(\sum_k T_{jk} f(A_k)\right)\right) \tag{7.13}$$

We can of course calculate the derivative of C with respect to any of the intermediate synaptic weights by applying the chain rule. The chain shown above demonstrates that the level of nesting of the parentheses, and consequently the complexity of the calculations, increases with the depth of the units. This is the disadvantage of this expression. However, it also has some advantages. If we calculate the derivatives, beginning with the output units and working backwards through the successive layers toward the input layer, most of the quantities necessary to calculate the derivative of a given layer have already been calculated for the other layers closer to the output. It therefore makes sense to use an iterative method, in which the partial derivatives are calculated in the direction opposite to the direction of propagation of the signal. This is where back-propagation gets its name.

The final version of the learning algorithm is as follows.

For each presentation h of a pattern s at the input:

1. *Phase 1. Calculation of the states:*
   For n increasing from 2 to p, the number of layers (propagation in the direction of the signal):
   The state S_i of each unit i is computed as follows:
$$S_i = f\left(\sum_j T_{ij} S_j\right) \tag{7.14}$$

2. *Phase 2. Calculation of the error gradients:*
   For each output unit:
$$\frac{\partial C}{\partial A_i} = \frac{\partial C}{\partial S_i}\frac{\partial S_i}{\partial A_i} = 2(S_i - Y_i)f'(A_i) \tag{7.15}$$
   For n decreasing from $p-1$ to 1 (back-propagation):
   For each cell j of the layer n:
$$\frac{\partial C}{\partial A_j} = \sum_i \frac{\partial C}{\partial A_i}\frac{\partial A_i}{\partial A_j} = f'(A_j)\sum_i \frac{\partial C}{\partial A_i} T_{ij} \tag{7.16}$$
   where the subscript i refers to the units receiving input from the unit j. (Put into words, this expression says that the influence of the unit j of the nth layer on the error C depends on the sum of the contributions to the error by the units i of the (nth + 1) layer, weighted by the influences of j on i (T_{ij}), modulated by the derivative of the sigmoid function at the point A_j.)
3. *Phase 3. Readjustment of the synaptic weights according to the gradient descent algorithm:*
$$T_{ij}^h = T_{ij}^{h-1} - 2\lambda^h \left(S_i - Y_i^h\right) f'(A_i) S_j \tag{7.17}$$

for the output layer and

$$T_{ij}^{h} = T_{ij}^{h-1} - \lambda^{h} \frac{\partial C}{\partial A_i} S_j \tag{7.18}$$

for the intermediate layers.

ARCHITECTURES AND PERFORMANCES OF LAYERED NETWORKS We have said that the architecture of the network depends on the problem to be solved. Unfortunately, no methods exist today to determine the architecture best suited for a given problem. We cannot even predict if an architecture is *a priori* capable of solving the stated problem. We could argue that an architecture which includes a large number of units distributed over multiple layers will perform better than a simple architecture, but this intuition is not always useful for the following reasons:

- the learning times grow very rapidly with the complexity of the network;
- a network which is too complex is not able to *generalize* from the examples given. To be more precise, the set of patterns to be classified is often so large that not all of them can be presented during learning. Only a subset of them, the training set, is presented. It follows then that during the recognition phase, the performance of the network is different for the training set and the other patterns. The ability to recognize these other patterns constitutes the capacity for generalization of the network. If the network is too complex, it will simply memorize the training set without generating an internal representation of the problem which would enable it to generalize.

In other words, we have an algorithm, but for a given problem we must:

- choose an architecture which is neither too simple, which would limit the nature of the correlations which the network can take into account, nor too complex, to limit the learning time and make generalization easier; and
- choose the best representation of the input patterns, and consequently, the structure of the input layer. The chosen representation is rarely the most compact, in order to make learning easier. Similarly, we must choose the representation of the equivalence classes for the output layer. Finally, for the hidden layers, we start by trying to learn with a minimum number of layers: a single one, or even no hidden layers. We then progressively increase the number of layers depending on the results obtained.

The choice of the number of presentations of each pattern depends on the successive changes in the T_{ij} observed during learning.

We first measure the performance of the network against the examples which were used for learning. We then test the generalization ability with patterns which the network has never seen.

The practitioner's know-how determines the success of an application, as it often does in computer science.

7–2–3 A Typical Application: NETtalk

Layered networks, whose potentials were first tested on simple problems, are now used in signal processing, in information coding, and in the recognition of acoustic patterns or images. One particular application, NETtalk, will serve as the basis for a description of some concrete problems, the architecture used, and the performance of a system.

NETtalk transcribes the letters of an English text into phonemes, or language elements. Using a standard encoding, the phonemes can then be sent to a specialized processor which supplies the input to a sound synthesizer. Learning takes place by presenting the letters of the text to the input layer and the appropriate phonemes at the output layer.

The architecture of the network is shown in figure 7.5.

Place encoding of the inputs is used, which means that there are as many units as different characters to code for. One unit corresponds to one character; it takes the value 1 if the letter is presented, and -1 otherwise. This method of encoding is therefore not very economical. In order to encode the 29 characters used, one uses

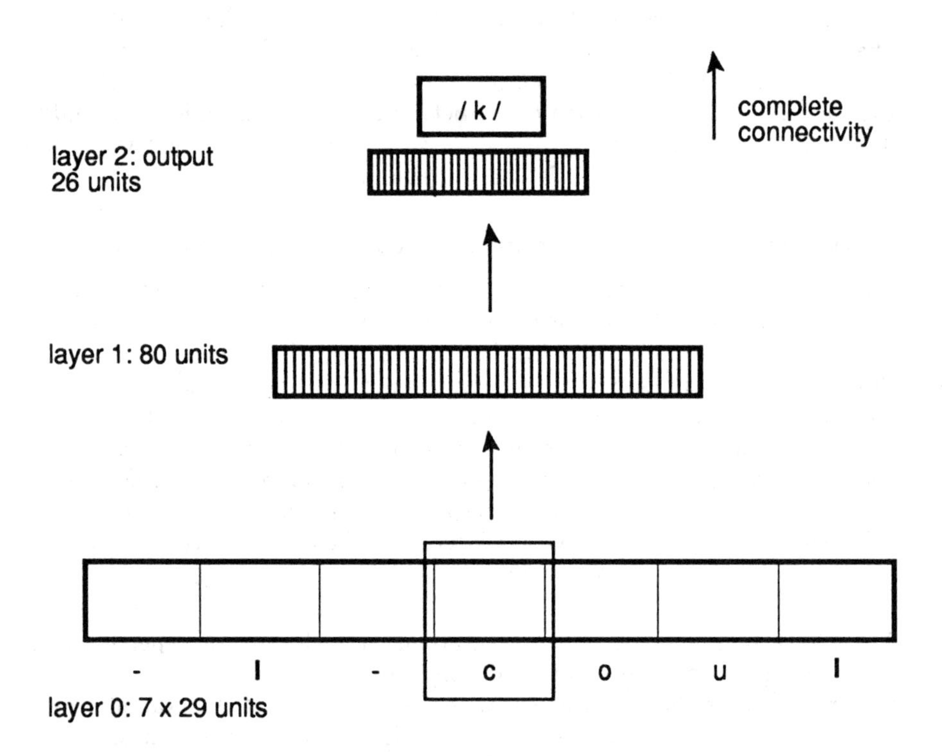

FIGURE 7.5 NETtalk, a layered network.

sets of 29 units. The input layer is composed of 7 such sets. The middle set receives as its input the letter whose phoneme must be determined, and the six neighboring sets receive the neighboring letters of the text. The difficulty of the problem lies in the fact that there is not a one-to-one correspondence between letters and phonemes, but rather it is context dependent. Three letters on each side of the letter read seem to be sufficient to take this effect into account.

A single intermediate layer contains 80 units connected to all of the input units.

The output layer also uses place encoding of the possible phonemes: it has 26 units.

PERFORMANCE The training set contains a thousand words. After the presentation of 50,000 words, the recognition rate reaches 95% for the training set, which produces quite an intelligible text. The performance drops to 78% when the words presented are not in the training set. The time needed for learning is equivalent to running a VAX 780 with floating processor all night.

THE ROLE OF THE HIDDEN UNITS One of the questions which is always asked about layered networks is the role of the hidden units. The states of the hidden units are related to the regularities which they detect in the patterns presented. In other words, they form an internal representation of the inputs. In the case of NETtalk, it would seem, as reported by T. Sejnowski and C. Rosenberg, that certain hidden units classify phonemes following classical phonetics rules, such as the distinction between vowels and consonants.

Conclusions

The architectures of most of the networks proposed for signal processing belong to one of two categories:

- Layered networks, in which the signal *propagates* during each time interval from one layer to the next. In these models, the layers are different and there are no feedback loops allowing the signal to go backwards. We have discussed the perceptron and networks based on back-propagation learning, but other models have also been proposed, in particular by K. Fukushima (the neocognitron) and B. Huberman (see bibliography).
- Homogeneous networks with feedback loops, in which the signal *iterates* until an attractive configuration is obtained. This is the case of the Hopfield networks and the many variations discussed in chapter 6.

Of course we could imagine that a combination of these two architectures might yield richer behaviors, but then new learning algorithms would need to be developed.

References

The first classic on the perceptron is the book by F. Rosenblatt, *Principles of Neurodynamics* (Spartan Books, New York, 1962). The second edition of the remarkably clear book by M. Minsky and S. Papert on the limitations of the perceptron has recently been published: M. Minsky and S. Papert, *Perceptrons* (MIT Press, Cambridge, 1988).

Layered networks are described at great length in the collective book *Parallel Distributed Processing*, edited by J. Rumelhart, J. McClelland, and the PDP Research Group, 2 vol. (MIT Press, 1986). The third volume, edited in 1987, contains much information about the algorithms, and includes a diskette for IBM compatible PC's. The chapter "Automata Networks and Artificial Intelligence" by F. Fogelman-Soulié, P. Gallinari, Y. Le Cun, and S. Thiria, in the collective book *Automata Networks in Computer Science*, edited by F. Fogelman-Soulié, Y. Robert, and M. Tchuente (Manchester University Press, 1987), compares linear formalisms and back-propagation, and addresses the problems of learning and generalization.

NETtalk is described by T. Sejnowski and C. Rosenberg in "Parallel Networks that Learn to Pronounce English Text," *Complex Systems*, vol. 1 (1987): 145–168.

Two other models of layered networks are described by K. Fukushima in "A Self-Organizing Multilayered Neural Network with a Function of Associative Memory: Feedback-Type Cognitron," *Systems Computers Controls*, vol. 6 (1975): 15–22; and by B. Huberman and T. Hogg in "Adaptation and Self Repair in Parallel Computing Structures," *Phys. Rev. Lett.*, vol. 52 (1984): 1048–1051.

EIGHT
Statistical Physics

The threshold networks described in chapter 5 can be generalized in several ways. One way consists of replacing the transition function by the probability of making a transition toward the different possible states. In this case we speak of probabilistic automata, which are mathematical objects whose behavior resembles that of particles in statistical physics. The quick overview that follows, and that constitutes the first part of this chapter, is not an introduction to statistical thermodynamics, but rather a restatement of some useful notions in the framework of the formalism of automata networks. This restatement will enable us to understand the connection between physical systems and automata networks, particularly in terms of numerical simulations. Both domains are enlightened at the conceptual and methodological levels. We will see that the memorization capacities of the Hopfield model can be "exactly" predicted using the statistical physics approach.

Statistical thermodynamics originated from the work begun at the end of the 19th century by L. Boltzmann and W. Gibbs, and continued by M. Planck, A. Einstein, P. Debye, and many others after them. Its objective is to describe the properties of macroscopic systems (i.e., systems at our scale), from the microscopic properties of their basic components. One starts from the simplest possible assumptions about the distribution of the components between the different microscopic states to which they have access. One can then use averages to obtain measurable physical quantities which characterize the macroscopic state of the system. The large number of components makes it possible to use averages rather than having to study all of the configurations which are in principle accessible to the system.

In this discussion we will only use quantum statistics, meaning that the states accessible to the particles are discrete. In fact, in the examples described, we will use Ising spins, denoted S_i, which are binary variables in the -1 or $+1$ states.

8–1 The Fundamental Quantities: Energy and Temperature

Physical systems are characterized by two features: the existence of an energy function and the role of the temperature.

8–1–1 Energy and Symmetry of Interactions

An *energy* function can be associated with each configuration. The existence of an energy function is a result of the internal symmetries of the system: the dynamical equations are invariant with respect to translation in time and the interactions between particles obey the principle of equal action and reaction. These symmetries only exist because the interactions are taken into account at the microscopic level. They can be absent in other models which use automata networks to model elements which are macroscopic objects. In particular, this is the case when these elements are living cells, which are macroscopic objects from the point of view of thermodynamics. Because of the analogy with physics, we easily get used to the idea that in very different domains, all dynamics evolve toward an attractor and that this attractor optimizes a function of the configuration of the system. These analogies range from economics (the maximization of utilities) to the evolution of species (maximization of fitness). In fact, as we have seen in chapter 2 of this book, the existence of attractors in no way implies the existence of an energy-type function which is optimized by the dynamics. *Chapters 5, 8, and 9 focus on the special case of systems for which there exists a function to optimize, such as the energy.*

In the case of spins, the energy is written in the form of a sum of terms:

$$E = \sum_i h_i S_i + \frac{1}{2} \sum_i \sum_j T_{ij} S_i S_j \tag{8.1}$$

where h_i is the local field applied to the spin i and T_{ij} is the interaction constant between spins S_i and S_j. If the sum is restricted to the terms which are linear in S, the particles are independent. The properties of the system and the mathematical methods are much simpler than those which we will mention for interacting systems, for which the quadratic terms are present.

8–1–2 Temperature and Probabilities

In general a physical system is not isolated from the rest of the world. Even if there is no external exchange of matter, there can be energy exchanges. This type of interaction is taken into account by the notion of *temperature*. With it comes the notion of probability. Though the properties of a physical system may be perfectly determined at the macroscopic level, they can fluctuate at the microscopic level. The probability of observing a configuration α of energy E_α is given by the Boltzmann ratio:

$$P_\alpha = \frac{\exp\left(-(E_\alpha/kT)\right)}{\sum_\alpha \exp\left(-(E_\alpha/kT)\right)} \tag{8.2}$$

where T is the temperature and k is a constant known as the Boltzmann constant. The sum in the denominator is performed over the set of all configurations accessible to the system.

At low temperatures, the arguments of the exponentials are large in absolute value, and the probabilities of low energy configurations are correspondingly much higher than the probabilities of high energy configurations. The higher the temperature, the larger the contribution of high energy configurations.

At zero temperature, the system evolves toward the energy minima. Its dynamics is reminiscent of that of the automata networks described in section 5.2: at a given instant, a particle "chooses" the state which lowers the energy or keeps it constant.

We also encounter a minimization principle at non-zero temperatures. The system evolves in such a way as to minimize a function, called the free energy, given by:

$$F = E - T \cdot S \tag{8.3}$$

where T is the temperature and where S, the entropy, is defined by

$$S = \sum_\alpha p_\alpha \cdot \log(p_\alpha) \tag{8.4}$$

where p_α is the probability of the α configuration with energy E_α. The more states are accessible to the system, the larger the entropy. Its influence also increases with temperature. At low temperatures the contribution of entropy is very small, and the minima of the free energy and the energy are the same. At high temperatures, the minimization of the free energy is obtained because the system occupies a large number of states with non-negligible probabilities, even if their energies are relatively high. In this sense the temperature favors disorder.

8–2 Probabilistic Automata and Monte Carlo Dynamics

Up to this point we have only mentioned static aspects, such as stable configurations of the dynamics. We might also want to describe the evolution of a system from an arbitrary initial configuration. The information about the energies of the different configurations is not sufficient to answer this question. This would require access to information about the probabilities of transitions between the different configurations of the system. Of course, when we can calculate them from physical principles, these transition probabilities depend on the temperature. In fact, if we are only interested in properties that solely depend on the distribution of the probabilities of occupation of the different configurations, it is sufficient to *simulate* the transition probabilities. The choice of transition probabilities is therefore dictated by the requirement that they generate stationary configuration probabilities that obey the Boltzmann law (Eq. 8.2).

One of the possible choices, the thermal bath method, is derived from a method proposed by Glauber. Consider a system composed of mutually interacting elements, denoted i. Let 1 be a configuration of energy E_1. We can calculate its probability of making a transition toward a neighboring configuration 2 with energy E_2, which differs from the preceding configuration only in the modification of the state of a single element i. We then choose a transition probability given by:

$$p_{12} = \frac{1}{1 + \exp(\Delta E/kT)} \tag{8.5}$$

where ΔE is the energy difference between the two configurations: $\Delta E = E_2 - E_1$ (Fig. 8.1).

The smaller the energy of the final state compared to the energy of the initial state, the larger the transition probability. This preference for lower energies becomes more pronounced as the temperature decreases. At zero temperature, the curve has an infinite slope, and we are back to deterministic dynamics: the system always evolves toward the lowest energy states.

In other words, taking the temperature into account in a dynamical system is reflected in the probabilistic nature of the changes of state of its components. These components are no longer described by deterministic automata, but by *probabilistic automata*.

A probabilistic automaton is defined by its sets of inputs and outputs, and by a transition probability toward each one of the possible output states as a function of the input configurations.

We can go from the transition function of the deterministic automaton to the transition probability of the probabilistic automaton by applying the following rule:

- In a random trial, the automaton has a probability p of going into the state calculated by the transition function, and a probability $1-p$ of going into the complementary state.

This leads to an idea for a numerical simulation method to find the low energy configurations of a physical system and, at the same time, to measure a number of physical quantities. We start with the system in a randomly chosen initial configuration. We choose at random an element i of the system to potentially undergo a change of state. We evaluate the energy difference between the two configurations 1, the initial state of the element i, and 2, the modified state of i. From it we deduce p_{12} by equation (8.5), and i undergoes the transition to state 2 with a probability p_{12}. We repeat this procedure, by choosing at random the element

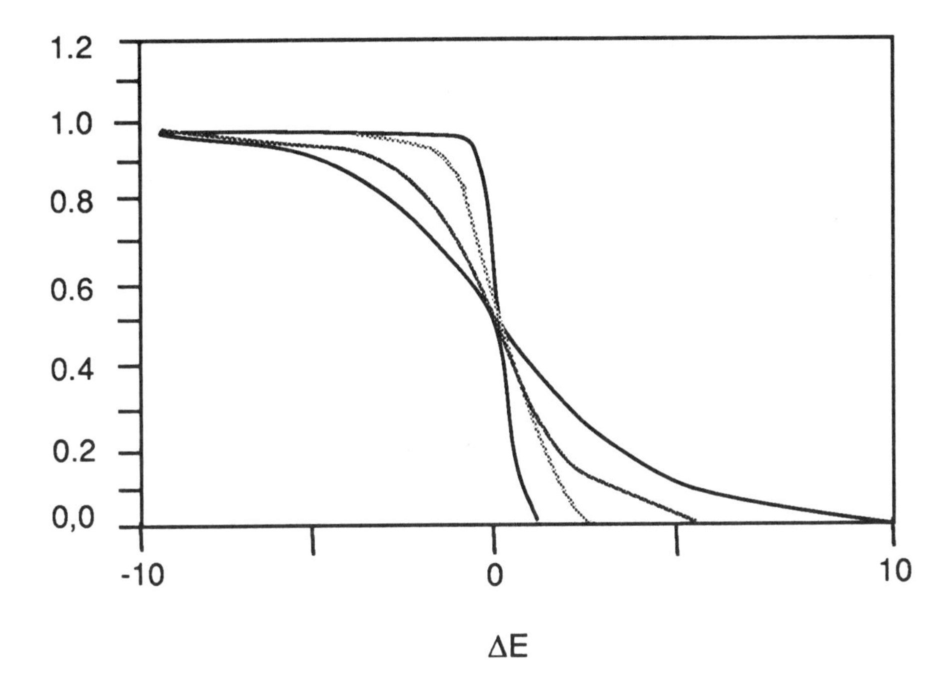

FIGURE 8.1 The transition probabilities between two neighboring configurations as a function of their energy difference ΔE for different values of the temperature T. kT, which is defined in the text, varies from 0.25 (for the most abrupt transition) to 0.5, 1, and 2 (for the most gradual transition).

which will potentially change its state, until on the average the properties which we have chosen to measure stabilize. This method is called the *Monte Carlo method*, because of the multiple random trials which it entails.

Before going on to the description of model magnetic systems, we will first explain their connection to automata networks. We are not interested in physical systems with a dynamical landscape, or an energy map, which reduces to a unique valley with a single minimum energy attractor, but rather in systems which have multiple valleys separated by barriers, similar in this respect to Hopfield networks where the numerous energy minima correspond to the references. The temperature now becomes an important parameter which controls the course of the dynamics. At low temperatures the system stays stuck at the bottom of a valley, as a function of the initial configurations; as the temperature increases, excursions of the system out of the bottoms of valleys become possible, until all of the configurations of the system become accessible.

8–3 The Ising Model

We will apply these considerations to magnetic properties of systems composed of elementary magnetic moments: spins. In the Ising model, the spins are placed at the nodes of a crystallin lattice which is usually square or cubic. Each element is therefore characterized by its spin, which takes on the value of +1 or −1, as a function of its interactions with its neighbors. To each pair of spins we associate an interaction energy:

$$E_{ij} = -T_{ij} S_i S_j \tag{8.6}$$

where T_{ij} depends on the pair of spins being considered.

The total energy of a configuration is given by:

$$E_T = -\frac{1}{2} \sum_i \sum_j T_{ij} S_i S_j \,. \tag{8.7}$$

The factor of 1/2 comes about because the sums are computed over all of the i's and all of the j's. The T_{ij}'s are of course symmetrical and the diagonal terms T_{ii} are zero.

At zero temperature, the dynamics of the spins is exactly as described in chapter 5 for threshold automata. The spins are updated by a random sequential iteration, and each one goes into the state which minimizes its energy of interaction with its neighbors. This amounts to comparing the sum of the energies of the pairs it forms with a threshold (zero, in this case) and taking on state +1 if this sum is negative when S_i is positive, and −1 in the opposite case. The attractors are indeed fixed points which minimize the energy function.

For non-zero temperatures, the behavior of the spin is no longer deterministic. It tends to "prefer" the minimum energy states, but the "thermal excitation" can send it toward higher energy states. Formally, the spin behaves like a probabilistic automaton. It makes a transition toward the state S_i with a probability given by:

$$P(S_i) = \frac{1}{1 + \exp[-(\sum_j T_{ij} S_i S_j / kT)]} \,. \tag{8.8}$$

In other words, when the sum of the $T_{ij} S_j$ is strongly positive (i.e. large compared to kT in absolute value), the spin S_i takes the value 1 with a probability which is close to 1. When the sum is strongly negative, this probability is very close to zero and the spin takes on the value −1.

In the indeterminate zone, of width $4kT$, the spin takes on one or the other value.

The dynamical properties of the magnetic medium depend on the choice of the T_{ij}, as was the case for networks of threshold automata. We have the choice of different connectivity schemes: cellular, complete, or random. The Ising model uses cellular connectivity. The T_{ij}'s can all be identical, or they can be different.

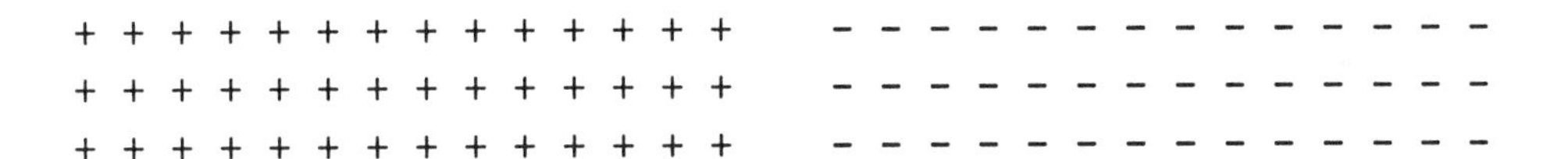

FIGURE 8.2 The two minimum energy configurations of a ferromagnet.

We will study three different cases: ferromagnetism (section 8.3.1), with identical interactions; spin glasses (section 8.5), with completely random interactions; and the probabilistic version of Hopfield networks (section 8.4), which is an intermediate case.

8–3–1 Ferromagnetism

QUALITATIVE DESCRIPTION We refer to the case in which the interaction constants are positive and identical ($T_{ij} = J$) as *ferromagnetism*. In the two configurations which minimize the energy, all of the spins are in the same state, $+1$ in one case and -1 in the other (Fig. 8.2).

At zero temperature, the system evolves toward one or the other attractor, depending on its initial configuration. As the temperature is increased, the probability of a spin leaving its energy well becomes non-zero, and from time to time certain spins go into a state which is not the same as that of most of their neighbors. Nonetheless, we can say that the configurations which are accessible to the system are localized in a valley around the minimum energy configuration. As the temperature is increased, more and more spins make transitions toward high-energy states, until the system is capable of going from one valley to another. For a system of infinite size, this phase transition happens at a well-defined temperature. The low-temperature phase, in which the system always stays in the same valley, is called the ferromagnetic phase. In the high-temperature phase, the system easily skips from one valley to another as time goes on, and the magnetization is zero on the average. This is called the paramagnetic phase.

THE MEAN FIELD TECHNIQUE As an example, we will now explain an analytic technique used to monitor the changes in the behavior of the system as a function of temperature. This approach, known as the mean field method, is but an approximation to the Ising model, but it can be broadly applied to much more difficult cases than the one which we will now study.

The probability that the spin i is in state E_i is given by equation (8.8). Suppose that in the sum over j of all of the interactions, we systematically replace S_j by its average value (averaged over all of the spins). This sum, $J \sum S_j$, is equivalent to a local magnetic field applied to the spin i by its neighbors j, which gives the mean field approximation its name. The average value of S_j, denoted $\langle S\rangle$, is also called the magnetization of the system. E_i, the energy of spin i, is given by:

$$E_i = -Jz\langle S\rangle S_i \tag{8.9}$$

where z is the number of neighbors of the spin i. This number is 4 for a square lattice, 6 for a cubic lattice, etc.

$\langle S\rangle$ is also the average value of S_i, which is given by:

$$\langle S\rangle = p_+ \cdot (+1) + p_- \cdot (-1) = p_+ - p_- \tag{8.10}$$

where p_+ and p_- are the probabilities that S_i is equal to $+1$ or -1, respectively. $\langle S\rangle$ is therefore the solution of the implicit equation:

$$\langle S\rangle = \frac{\exp(zJ\langle S\rangle/kT) - \exp(-zJ\langle S\rangle/kT)}{\exp(zJ\langle S\rangle/kT) + \exp(-zJ\langle S\rangle/kT)} = \tanh\left(\frac{zJ\langle S\rangle}{kT}\right) \tag{8.11}$$

The number of roots of the implicit equation in $\langle S\rangle$ depends on the slope at the origin of the hyperbolic tangent function compared to 1 (Fig. 8.3). This slope is inversely proportional to the temperature. For small slopes, the line of slope 1 intersects the graph only at $\langle S\rangle = 0$. It follows that the magnetization is zero at high temperatures. At low temperatures, on the other hand, the slope is high and the line intersects the graph at three points. The point of zero magnetization is in fact unstable, for it corresponds to a free energy maximum, as would show the direct application of the equations given above (equations (8.3) and (8.4)). The two other intersections correspond to stable configurations. The lower the temperature, the closer the two opposing magnetizations are to $+1$ and -1. T_c, the temperature

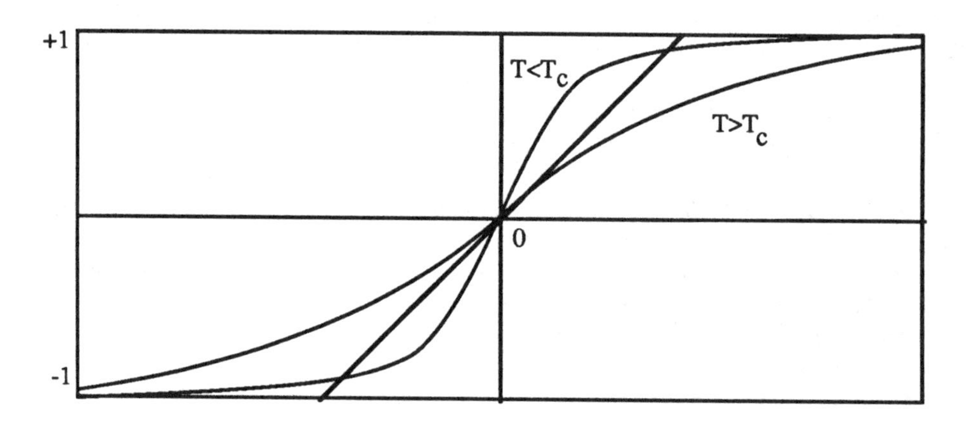

FIGURE 8.3 Graphical solution of the implicit magnetization equation (8.11).

of the transition between the two behaviors, paramagnetic at high temperatures and ferromagnetic at low temperatures, is given by:

$$T_c = \frac{zJ}{k}. \tag{8.12}$$

$\langle S \rangle$ acts as an *order parameter*, zero in the disordered high-temperature phase, and of macroscopic amplitude in the ordered low-temperature phase. It takes on a maximum amplitude of 1 at zero temperature: the order is then maximal, since only a single configuration at a time is accessible to the system.

The conclusions of this approximate calculation are still valid when we use more sophisticated methods. They can also be verified experimentally. To summarize:

- At high temperatures, the configurations of zero magnetization are accessible. The system evolves from one to the other, since the transition probabilities between configurations with similar energies are very high. We say that the system is in a unique *isotropic phase*. The order parameter is zero.
- At low temperatures, two phases of non-zero magnetization can exist and as a consequence of the dynamics, the system "chooses" one of them as a function of the initial conditions (and of the external magnetic field, which can favor one of the two phases, in a physical system). Each phase is stable: a system of infinite size has zero probability of going from one phase to the other. The symmetry of the system between the +1 and −1 magnetization configurations is therefore broken at low temperatures. The order parameter is non-zero.
- The transition between the two behaviors happens at a critical temperature which is proportional to the interaction.

8–3–2 Generalizations

The physics of magnetic media leads to several simple generalizations of ferromagnetism.

The interactions between neighboring spins can be uniform and negative. In this case the medium is known as an antiferromagnet. There are two opposing minimum energy configurations: in each of them, the spins alternate between positive and negative values, going from one spin to its nearest neighbor (Fig. 8.4).

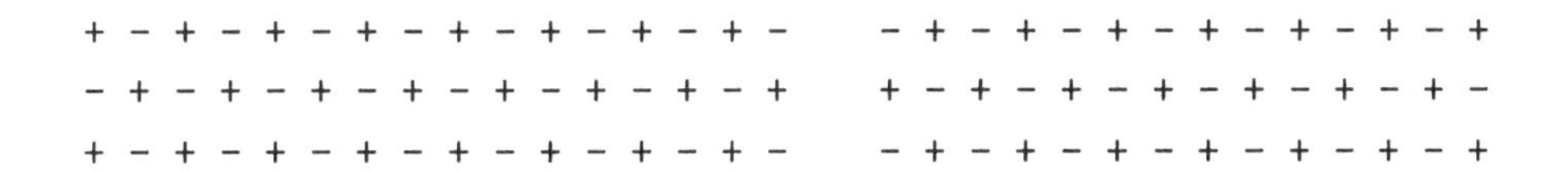

FIGURE 8.4 The two minimum energy configurations of an antiferromagnet. They are equivalent to within a translation.

Although the net magnetization is zero, the system is perfectly ordered. We also see two phases depending on the temperature, separated by a phase transition. In the case of interactions with the second nearest neighbors, or of anisotropic interactions, we can once again see behaviors characterized by a phase transition between a disordered high-temperature phase and ordered low-temperature phases.

Up to this point, we have only considered uniform interactions throughout the lattice. Introducing some disorder can complicate matters. However, some kinds of disorder only lead to behaviors with two attractors at low temperatures. If all of the interactions remain positive, they always favor the alignment of the spins, and the two stable configurations are the magnetization configurations near +1 and −1.

The same is true of the Mattis model, in which all of the interactions are of the form:

$$T_{ij} = U_i U_j \tag{8.13}$$

where the variables U_i, randomly distributed, are either +1 or −1, which is similar to a Hopfield system with a single reference. In fact, the change of variables:

$$S_i' = U_i S_i \tag{8.14}$$

is sufficient to make the new interaction constant between the S'_i equal to 1, and consequently independent of the pair of neighboring spins being considered. It follows that at low temperatures, we again find two attractors near $S'_i = +1$ and $S'_i = -1$, corresponding to $S_i = U_i$ and $S_i = -U_i$, respectively. Of course the magnetization, which is related to the variables S_i, has no reason to be near +1 or −1. The order parameter is now the projection of the S_i onto the U_i:

$$q = \frac{1}{N} \sum_i S_i \cdot U_i \,. \tag{8.15}$$

Depending on whether $S'_i = \pm 1$, q will be near +1 or −1.

8–3–3 The Frustration

In fact, a "mere" disorder is neither necessary nor sufficient to introduce multiple attractors. In addition, the interaction loops must be *frustrated*. This notion of frustration is the same as the one we introduced in chapter 2 for automata networks with a single input. In this case, the frustration of a loop is evaluated by computing the product of the interactions of the connections around the loop.

If this product is positive, the loop is not frustrated. It is possible to find two opposite configurations which are absolute minima of the loop energy (Fig. 8.5).

If the product of the interactions is negative, the loop is frustrated and it is not possible to find a configuration which satisfies all of the interactions. Instead, there exist several configurations (a total of eight for interactions equal to +1 or −1 such as in figures 8.5 and 8.6) which satisfy three out of the four interactions. Their energy is therefore −2 (Fig. 8.6).

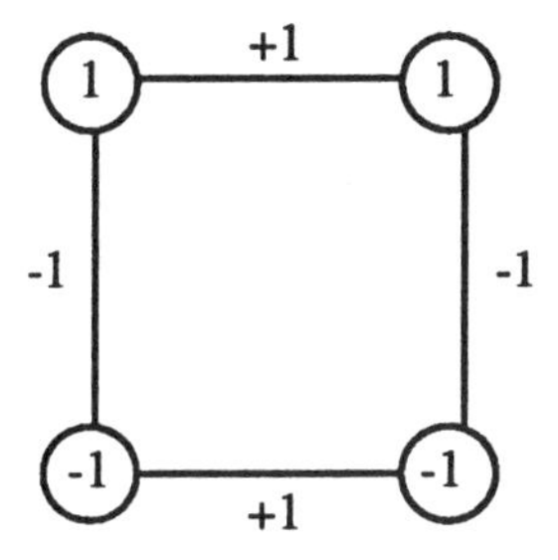

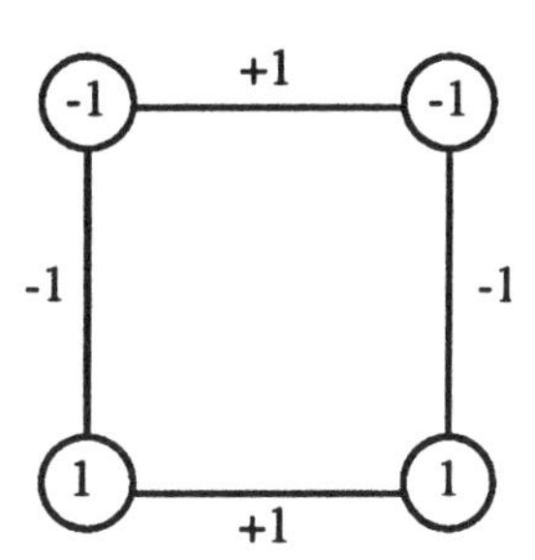

FIGURE 8.5 Two minimum energy configurations of an unfrustrated loop. The numbers next to the connections are the interaction constants, and those at the four corners are the states of the spins. For both configurations, the energy of each connection is indeed negative and equal to –1. The energy of the loop is therefore –4 in both cases.

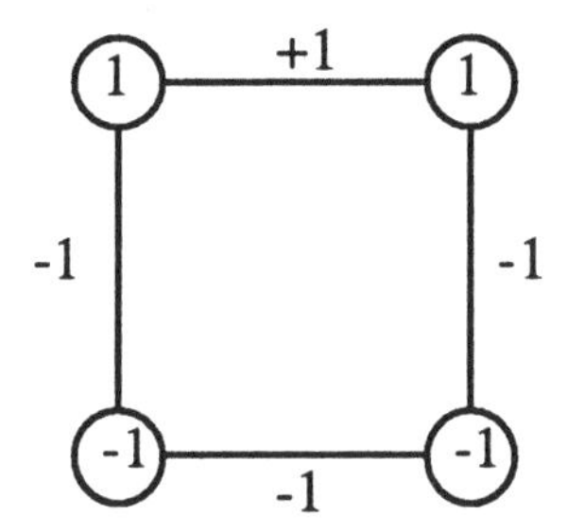

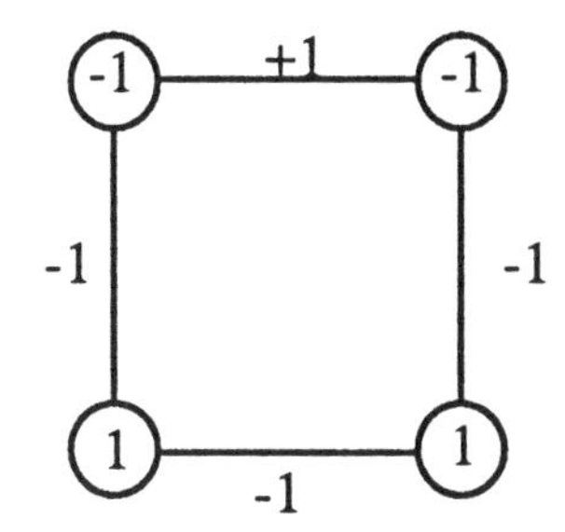

FIGURE 8.6 Two configurations of a frustrated loop. The energy of both loops is –2. No configuration can simultaneously minimize the energies of each of the four connections; there exist six other equivalent energy configurations.

In the case of a system in which there are no frustrated loops, such as in the two examples shown above, the order induced on one loop propagates from loop to loop. On the other hand, in the case of a frustrated system, i.e., a system which has many frustrated loops, the frustrated areas can coexist in many configurations relatively independently, which makes it possible to have a large number of attractors.

8–4 Formal Probabilistic Neurons

The behavior of networks of formal probabilistic neurons is intermediate between that of the ferromagnetic systems described above and spin glasses, for which the interaction constants are chosen completely randomly. The questions arising from this generalization of systems of formal neurons were studied by a group of researchers, the main contributors being D. Amit, H. Gutfreund, and H. Sompolinsky.

This model involves spins whose interactions, as in the Hopfield model, are given by:

$$T_{ij} = \frac{1}{N} \sum_{s} S_i^s \cdot S_j^s \tag{8.16}$$

where the $1/N$ factor normalizes the strength of the interactions so as to keep the results independent of the number of automata, when this number goes to infinity. The S^s are the reference configurations. This is a fully connected system, in contrast to the Ising model, where the interactions only involve the nearest neighbors.

Recall that the probability that automaton i goes into state S_i is given by equation (8.8), in which the synaptic weights are those of equation (8.16).

This probability function of S_i corresponds to the energy function shown in figure 8.1. The temperature plays the same role of widening the zone of uncertainty. In particular, at zero temperature $P(S_i)$ degenerates into a Heaviside step function, and the probabilistic model reduces to the deterministic dynamics of the Hopfield model studied in chapter 5.

As in the case of the Hopfield model, we are interested in the associative memory properties, and therefore in the attractors of the dynamics defined by the network of T_{ij}'s calculated as a function of the references using the Hebb rule. A random sequential type of iteration is used. The *stricto sensu* attractors defined in the case of deterministic automata must be generalized. In the case of probabilistic dynamics, we speak of an attractor if the system evolves around an attractive configuration, staying within a Hamming distance which is small compared to N. Finally, as long as the interactions are symmetric and the references are random and uncorrelated, the two parameters which define the dynamical behavior are the temperature and α, the ratio of the number of references to the total number of automata.

We can of course use numerical simulations to study the performance of such a system. The simulation results agree with the results of the formal study using thermodynamical methods which we will now describe.

Let's begin by defining m quantities which we use as order parameters. These are the projections of the attractive configurations S^a onto the references S^s:

$$q_{as} = \frac{1}{N} \sum_{i} S_i^a \cdot S_i^s \tag{8.17}$$

In other words, instead of explicitly calculating the attractive configurations, we will simply calculate their projections onto the references. An attractor which is near a reference S^μ will have projections onto the other references which will be close to zero (of the order of $1/\sqrt{N}$); the projection onto S^μ will be equal to 1.

$$q_{as} = \delta(a, s) \tag{8.18}$$

where δ is the Kronecker delta function, which is equal to 0 if a and s are different, and 1 if they are the same.

The study done by D. Amit, H. Gutfreund, and H. Sompolinsky is based on the existence of mean field equations, analogous to the equation which gives the magnetization in the case of ferromagnetism, which can be used to calculate the projections q_{as}. Since the connectivity is infinite, these equations are exact. Different types of solutions which have non-zero projections q_{as} appear at temperatures less than 1.

The most interesting solutions are very close to the reference patterns. In fact, for each of them there exists a corresponding opposite configuration. However, they are only equivalent to the reference patterns if the temperature is zero and α goes to zero, which is in agreement with the results obtained in chapter 5. Nonetheless, if we tolerate a very small percentage of errors, of the order of 1%, we observe attractors for which a single projection onto the references is close to 1, and all the other projections are negligible, as long as the number of attractors is less than about 14% of the number of automata. However, there exists a critical fraction α_c above which the projections q_{as} onto the attractors go to zero in a discontinuous fashion. We can get an idea of the accuracy of the recognition of the references by examining the variation of the error percentage e as a function of α at zero temperature. e is related to the projection q_{as} as follows:

$$e = 100\frac{1 - q_{as}}{2}. \tag{8.19}$$

We can see that e goes up very slowly as α approaches α_c, and that it always remains less than 1%. As soon as α_c is exceeded, e immediately jumps to 50%, which corresponds to uncorrelated configurations (Fig. 8.7).

We also note the appearance of "parasitic" solutions formed from combinations of several references for the lowest temperatures and fractions α. Figure 5.3.e showed an example of such a solution. Only combinations of an odd number of references are stable, but always less so than the attractors near the references. These combinations are obtained by adding the spins of each reference at each node: if the sum is positive, the node takes on the value 1; otherwise it takes on the value -1. The number of parasitic solutions is much higher than the number of references: it is exponential in m.

In fact, raising the temperature, or increasing α is equivalent to adding noise to the dynamics of the convergence toward the attractors, which eliminates the most unstable ones. We can then draw a true phase diagram as a function of the two parameters α and T, in which curves such as T_M and T_3 separate phases of different natures. Above the T_M curve we observe a spin glass phase, in which coexist a very

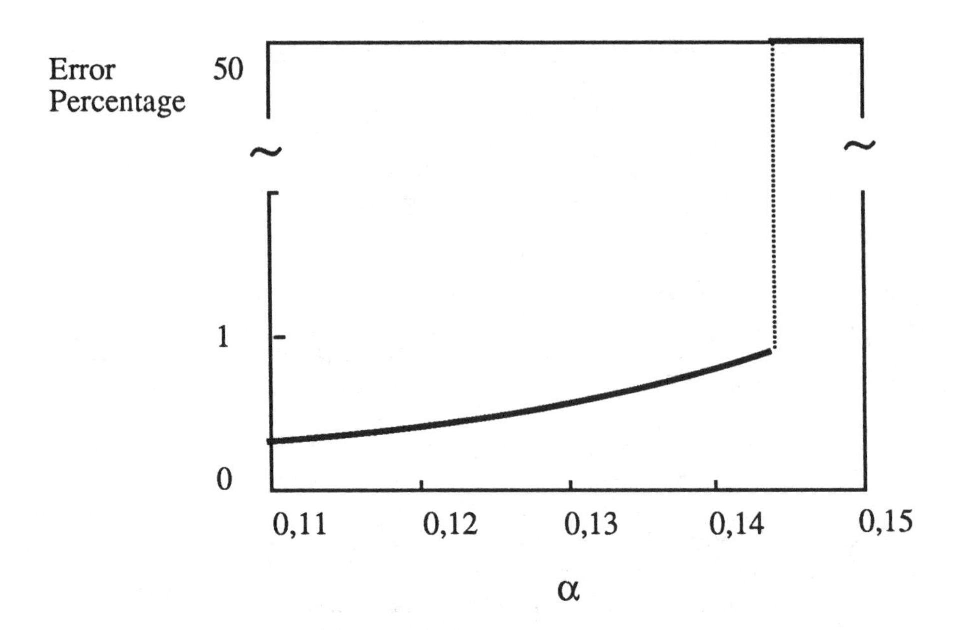

FIGURE 8.7 Accuracy of recognition at zero temperature. The error percentage between the attractor and the nearest reference stays small until the "catastrophe." Afterwards, the system completely loses its memory. (From D. Amit, H. Gutfreund, and H. Sompolinsky, *Annals of Physics*, vol. 173 (1987): 30.)

large number of attractors which have small (of the order of $1/\sqrt{N}$) projections onto all of the references. Under the T_M curve is a phase in which configurations very close to the references are attractive. These are not, however, the only attractors. Finally, if the temperature drops below T_3, the combinations of three references become stable. We can therefore consider that a certain level of noise, represented here by a non-zero temperature, can improve the performance of the system as an associative memory by destabilizing the combinations of references which we would like to avoid. This noise exists for real neurons, and is caused by physical and chemical processes at the level of the membrane. Note that thermal noise and noise due to the presence of other references both have the same effect of destabilizing the reference (Fig. 8.8).

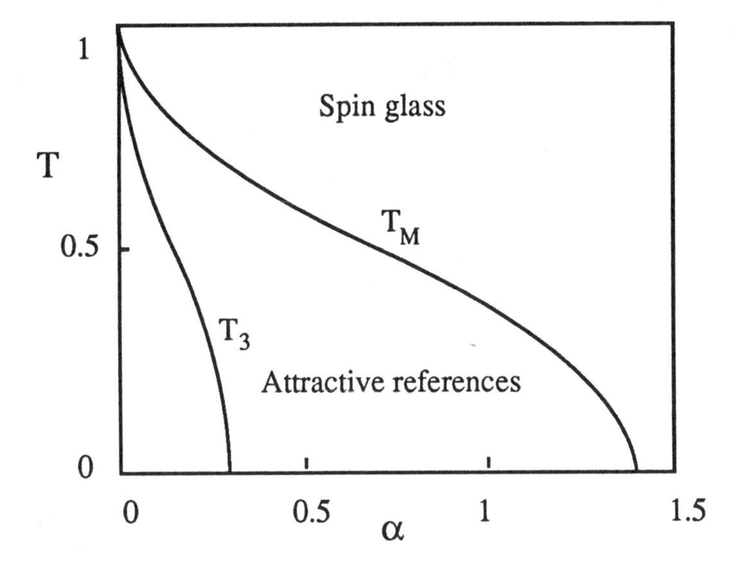

FIGURE 8.8 Phase diagram of networks of formal probabilistic neurons. We can distinguish three different phases: at high temperatures and a large number of references, a spin glass phase. In the intermediate phase, the references are attractive. At low temperatures and small number of references, combinations of references are also stable. (From D. Amit, H. Gutfreund, and H. Sompolinsky, *Annals of Physics, ibid.*)

8–5 Spin Glasses

Spin glasses are a model of glasses in general, and of diluted magnetic media in particular. Solids can exist in two extreme forms: the crystalline, or ordered form, which corresponds to a regular arrangement of atoms, and the amorphous, or disordered form, which is characteristic of glasses. The irregular arrangement of atoms translates macroscopically into different thermodynamic properties. For example, the liquid-solid phase transition, known as the vitreous phase transition for amorphous solids, does not have the abrupt character observed in the case of crystals. The transition temperature can vary if the cooling is too rapid, and there are numerous hysteretic effects. Rather than work on glasses in which the structural disorder is difficult to control, physicists often work on a simpler model system, that of diluted magnetic media, since their composition can be controlled and a large number of physical measurements are possible. The disorder is then in the magnetic interactions between the spins.

Spin glasses have been the subject of numerous studies, both experimental and theoretical, and so we will not discuss anything that relates directly to magnetism or physics proper. Instead, we will set out to describe the metastable states of the dynamics, for the sake of comparisons with the other systems studied in this book.

The Sherrington-Kirkpatrick model is a system in which each spin interacts with all of the others in a fixed but random way, independently of distance. This simplification justifies the mean field method, and allows the thermodynamic properties to be calculated using the replica method, which we will not discuss here. The results can be summarized as follows:

The high-temperature phase, or the paramagnetic phase, is comparable in all respects to the paramagnetic phase of the simpler systems described earlier: the system has access to the different configurations with a finite probability.

A spin-glass phase appears at low temperatures, with attractors characterized by the following properties:

- The number of attractors is exponential in the number of spins. It has been shown to be of the order of $\exp(0.2N)$ at zero temperature.
- In order to characterize the ordered nature of the spin-glass phase, we would like to determine an order parameter by projecting the attractors onto one or several reference configurations. As a result of the disorder of the interactions, this is not possible, and we must instead look at the crossed

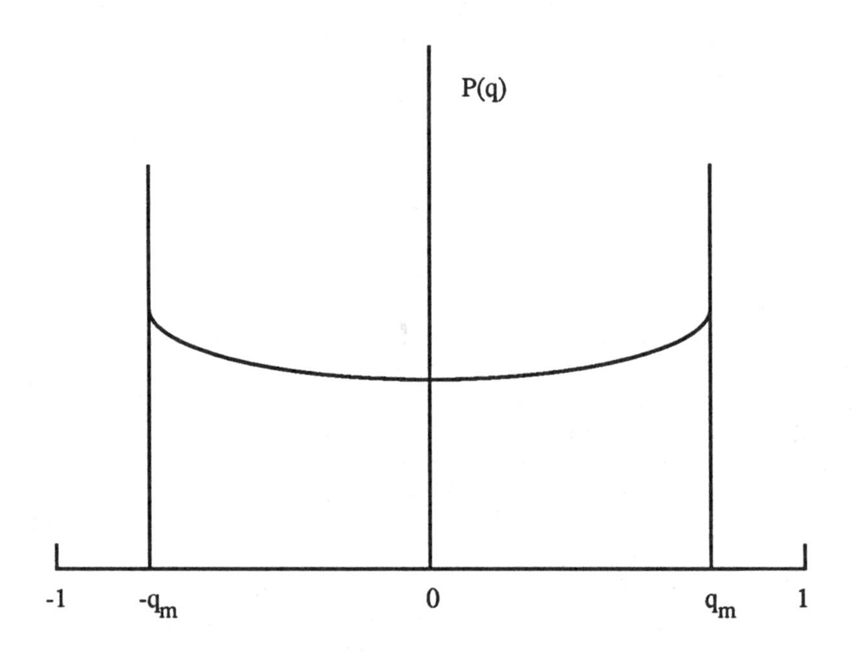

FIGURE 8.9 $P(q)$, the histogram of the projections between the attractors for a spin glass.

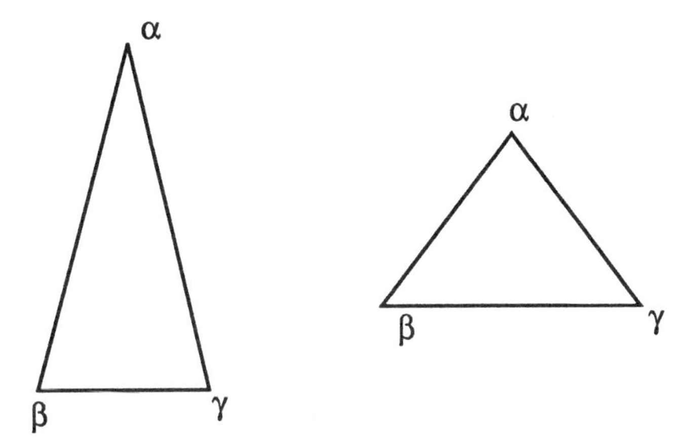

FIGURE 8.10 The characteristic triangles of ultrametrics.

projections between the attractors. We define $q_{\alpha\beta}$, the projection of the two attractors α and β, by:

$$q_{\alpha\beta} = \frac{1}{N} \sum_i S_i^\alpha \cdot S_i^\beta \,. \tag{8.20}$$

Instead of a single order parameter, as in the case of ferromagnetism, or of a finite number of parameters, as in the case of a formal network of neurons, we must define an order function, which is the histogram of the $q_{\alpha\beta}$:

$$P(q) = \sum_{\alpha\beta} p_\alpha p_\beta \delta(q_{\alpha\beta} - q) \,. \tag{8.21}$$

$P(q)$ is the probability that the projection of two configurations α and β onto each other is q. p_α and p_β are the probabilities of α and β, respectively, and δ is the Dirac delta function. The shape of $P(q)$ is shown in figure 8.9.

We obtain two Dirac delta functions at $-q_m$ and q_m and a continuum of possible projections in between. (At high temperatures, $P(q)$ would have only a single delta function at $q = 0$.)

- The distribution of the attractors, however, is not arbitrary. If we consider triplets of attractors $\alpha\beta\gamma$, their respective Hamming distances form either equilateral triangles or isosceles triangles whose two equal sides are always longer than the third side (Fig 8.10).

This property, which defines *ultrametrics*, is characteristic of hierarchical systems.

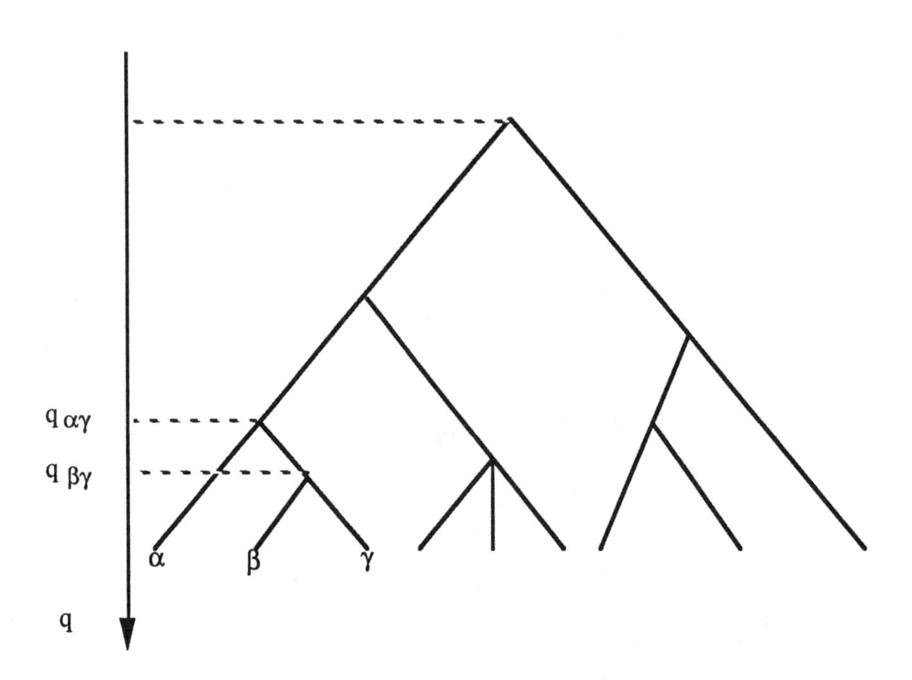

FIGURE 8.11 The hierarchical structure of attractors in a spin glass.

The structure of the attractors is a hierarchical tree structure, in which the attractors are represented by the leaves of the tree, and their respective distances by the height of their first common root (Fig 8.11).

We can see here the origin of the triangle rule stated above.

- Finally, despite the large number of attractors, only a small number of them have a large basin of attraction, and consequently a large statistical weight. The basins of these attractors occupy a finite percentage of the state space. At low temperatures, the properties of the system are due to one of these large basins, provided the system was cooled sufficiently slowly.

All of the properties stated above were obtained for the Sherrington-Kirkpatrick model. Whether these properties hold in real physical three-dimensional systems is still an open question.

References

The book by C. Kittel and H. Kroemer, *Thermal Physics* (Freeman, 1980) is a simple introduction to thermodynamics.

The most important papers on the subjects touched on in this chapter are collected by M. Mézard, G. Parisi, and M. Virasoro, in the book *Spin Glass Theory and Beyond* (World Scientific, 1987).

NINE

Simulated Annealing

In this chapter we will study a combinatorial optimization method: simulated annealing. Optimization problems which occur in technological applications can quickly become forbiddingly complex and require huge computation times because of the combinatorial explosion of the number of possible solutions. A simple example, the task assignment problem, will serve to illustrate some of these difficulties.

The Assignment Problem

Suppose that a business would like to distribute N different tasks (related to manufacturing, the delivery of raw materials, shipping the finished products, accounting, etc.) among N different locations of a building. Each distribution has some advantages and some disadvantages with respect to the appropriateness of the task to the location, or to the problem of the transfer of materials or information between the locations. Suppose in addition that we can quantify, or at least classify, the efficiency of each distribution, for example in terms of the price of the manufactured products. The task assignment problem is that of determining the most efficient distribution.

The brute force method consists of calculating and comparing the efficiencies of the $N!$ possible distributions in order to determine the best one. In other words, the complexity of the problem in terms of the computation time needed to solve it increases at least as fast as $N!$. These exponential, or non-polynomial, computation times rule out "brute force" algorithms as soon as the number of components is greater than a few dozen.

Another approach which can be used is the steepest descent method, which is derived from the gradient descent method and is less computationally intensive than the brute force method. It consists of starting from an arbitrary initial distribution and randomly trying permutations. For example, we exchange the locations of two tasks and compare the new efficiency to the old one. If the efficiency has increased, the new distribution is temporarily adopted and used as a new starting point for a new trial. If the efficiency decreases in a given trial, we return to the previous configuration and try a different permutation. The algorithm continues until, after a large number of consecutive negative trials, it does not seem possible to improve on a distribution.

This algorithm functions more or less well, depending on the structure of the efficiency function. It works well if this function is convex over the set of distributions, in other words, if there exists only one global optimum.

The difficulty comes from the potential existence of several local optimum distributions. A local optimum is defined by the fact that any partial permutation—in a predetermined set, such as the set of pairwise permutations—decreases the efficiency function. The efficiencies of the local optima can be very different. Yet, the initial problem was to obtain the best efficiency among all of the different possible distributions, and consequently among the local optima. When there are several optima, the gradient descent method can result in the system getting "stuck" in a configuration which is not the best one. One variation of this method is to try different initial configurations, chosen randomly, or cleverly when we can; to compare the local optima attained; and to choose the best one. But there is no guarantee that one of the trials has reached the global optimum, especially if there are many local optima. This leads to the idea of choosing "good orientations" for the initial configurations of each trial.

Thermal Annealing

Let's return to thermodynamics. The "search" for the lowest energy states by a physical system is formally analogous to a combinatorial optimization process. At zero temperature, the system evolves from a given configuration toward the lowest energy configuration or configurations by a gradient-descent-type method. Consequently, in the case of a frustrated system, the configuration reached is frequently a metastable state, which has an energy greater than the absolute minimum. The system is in some sense trapped in this local minimum (Fig. 9.1).

On the other hand, at a non-zero temperature, the probabilistic nature of the configuration changes can allow the system to go uphill and to leave the basin of attraction of one relative minimum to go into another, deeper basin. Ideally, in order to reach the minimum, the temperature would have to be high enough to

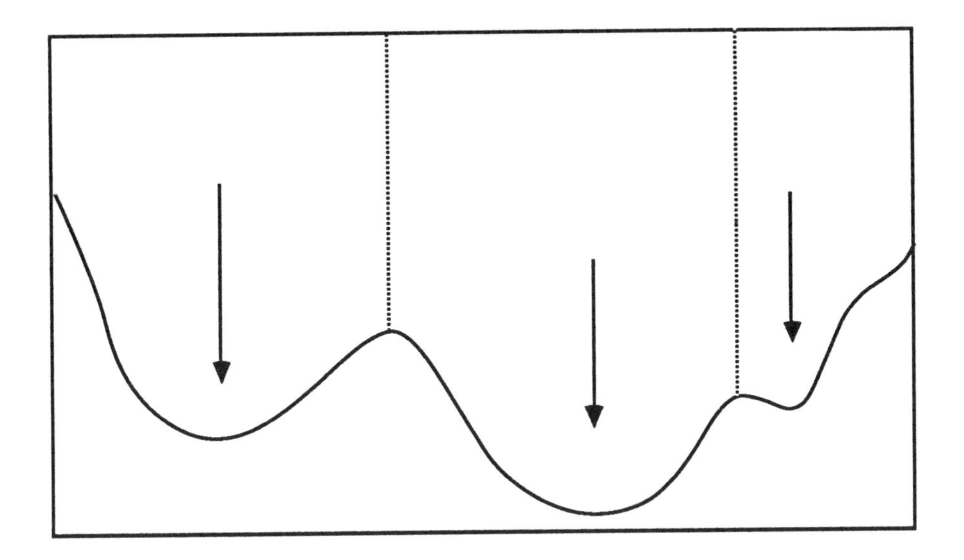

FIGURE 9.1 An energy landscape. Depending on the initial configuration, represented by the arrows, at zero temperature the dynamics ends up in one of the relative minima, separated by the barriers represented by dotted lines. At high temperatures, the probabilistic processes allow the system to jump over the barriers which separate the valleys.

enable the system to jump over the barriers, but low enough to nonetheless attract it to the bottoms of the valleys.

Consider the following double dynamics: first a search for minima at a fixed temperature, followed by a dynamical decrease of the temperature. If we start at a high temperature, all of the configurations are accessible and the system has only a weak preference for lower energy states. By progressively lowering the temperature, we allow the system to look for relatively large basins of attraction initially, while keeping it from being trapped by metastable attractors. The number of accessible configurations depends on the temperature. At average temperatures, the system seems to be sensitive only to the coarsest features of the dynamical landscape. By slowly lowering the temperature, we keep the system from getting trapped in high-energy valleys, in order to send it to the largest, deepest basins, where subsequently lowering the temperature will send it to the bottom.

Such a mechanism is known as *annealing*, by analogy with the thermal cycles used in metallurgy. When a metal or alloy is cooled, the rate at which the temperature is decreased is chosen based on the desired mechanical properties of the solid. The steel used to make cutting tools, for example, is quenched by cooling it very fast. This operation yields a very imperfect crystal in a metastable state, but very hard. On the other hand, if we would like to obtain a flexible material which

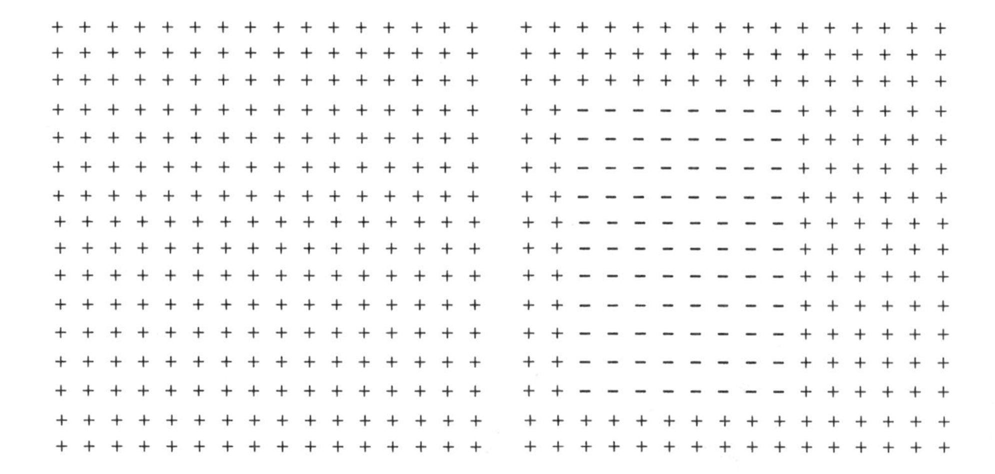

FIGURE 9.2 Low energy configurations of a ferromagnet. On the left is shown the minimum energy configuration, and on the right a metastable configuration, which at non-zero temperatures can potentially evolve toward the configuration on the left.

is not brittle, we need to make the crystalline structure as perfect as possible. In this case the annealing is done at intermediate temperatures in order to eliminate the defects of the crystal. Figure 9.2 illustrates these phenomena in the case of a ferromagnet.

The homogeneous structure on the left of figure 9.2 is obtained at zero temperature using annealing: it corresponds to a minimum energy configuration. The structure on the right is obtained by quenching. We can see that within the same structure, domains coexist with both positive and negative magnetizations. If these domains were isolated, each one would correspond to an energy minimum. However, the boundaries between the domains add a positive contribution to the energy. Eliminating them would require the simultaneous switching of a large number of spins, and consequently a momentary increase of the energy of the system, which is very improbable at this low temperature. The system is stuck in a metastable state.

The thermal properties of glasses are related to the existence of their multiple attractors. If the glass is cooled quickly, the atomic configuration is metastable; it is obtained for a relatively high temperature, which we will call T_r. On the other hand, if we anneal the material, the system has time to change its configuration between different metastable states, and so the system continues to evolve, even for temperatures less than T_r. The transition temperature is therefore lower when cooling is done slowly. In the case of silica, depending on the rate of cooling, we obtain either an amorphous form commonly known as silica, or a quartz crystal.

9-1 Simulated Annealing

The idea behind simulated annealing in combinatorial optimization is to numerically simulate a thermal annealing operation. The principle is as follows:

Consider a system composed of N elements. To each configuration of the system is associated a function to optimize. If we want to find the configuration which minimizes this function, it acts as the energy. If instead we look for a maximum, we take as the energy the function of opposite sign. The configurations of the system can be modified by discrete state changes of the components of the system. Of course, the systems which we are interested in are those with a large number of optima. We start with a random configuration (or a configuration chosen carefully as a function of the problem). At the start of the iteration, we set a temperature parameter according to the range of energies accessible to the system. We then iterate the following process (known as a Metropolis or Monte Carlo process): we randomly modify the current configuration, which changes the energy of the system by ΔE. If the energy decreases, we retain the modification. If the energy increases, we retain the modification with a probability given by $\exp(-\Delta E/T)$. (The Monte Carlo and probabilistic automata processes are equivalent with respect to the stationary states obtained. Probabilistic automata require a few more calculations of exponentials and random trials, and so for practical applications the Monte Carlo method is generally preferred.)

We continue to iterate as long as the energy of the system is decreasing. When the energy stops changing, we reduce the temperature and continue the Monte Carlo process of reducing the energy. The process continues until further reductions in the temperature are ineffective.

We will use three problems chosen in three different fields to illustrate this method: the traveling salesman problem (section 9.1.1), the partition problem (section 9.1.2), and image processing (section 9.2). The first problem is a classical one, albeit somewhat academic.

9-1-1 The Traveling Salesman Problem

This is the quintessential combinatorial optimization problem. A number of cities are distributed on a map and the traveling salesman must organize a tour which visits each city exactly once before returning to the starting point. Of course the object is to find the shortest tour. If we translate the problem into the more general language developed earlier, we associate with each city a number between 1 and N corresponding to its order in the tour, where N is the total number of cities. The energy of a permutation is the total length of the path determined by the tour.

The initial configuration is not generally chosen at random. One effective possibility is to start from the first city and go to the nearest city. We continue this algorithm, going at each step to the nearest city which has not yet been visited. After reaching the last city, we return to the starting city.

The elementary permutation of a trial consists of inverting a section of the tour. For a given tour, we permute four cities as shown in figure 9.3.

Other permutations, involving six cities, for example, are possible and even more effective.

Figure 9.4, taken from the original paper by Kirkpatrick et al., shows the results of such a simulation for a set of 400 cities distributed inside a square. The length of a side of the square is set equal to $\sqrt{N}$ so that the average distance between two neighboring cities is on the average independent of N. In this framework, the energy of a tour varies as αN, where α is a dimensionless quantity of order 1. The temperature ranges from $\sqrt{N}$ to 1. Each of the four tours represents a stationary

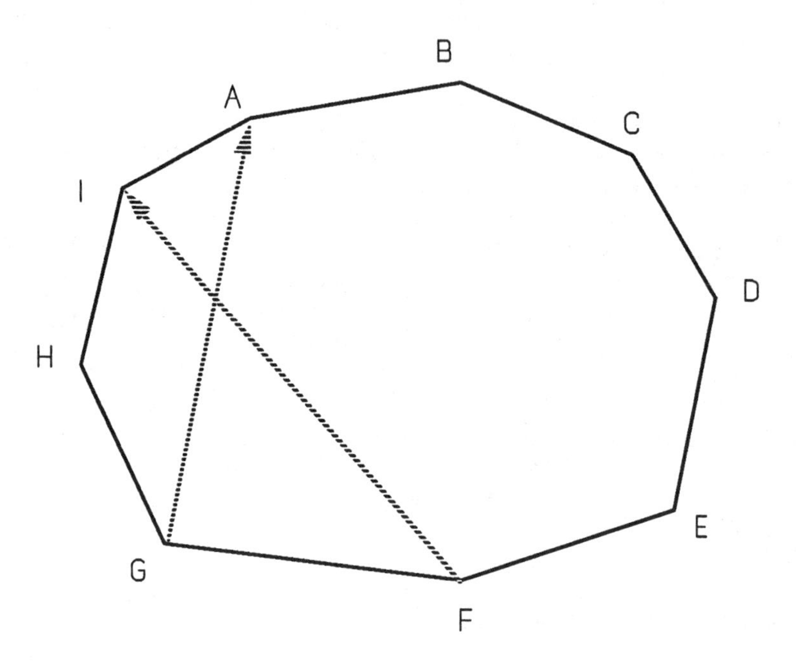

FIGURE 9.3 The permutation shown on the figure changes the tour from ABCDEFGHIA to ABCDEFIHGA. (It is not a great success!)

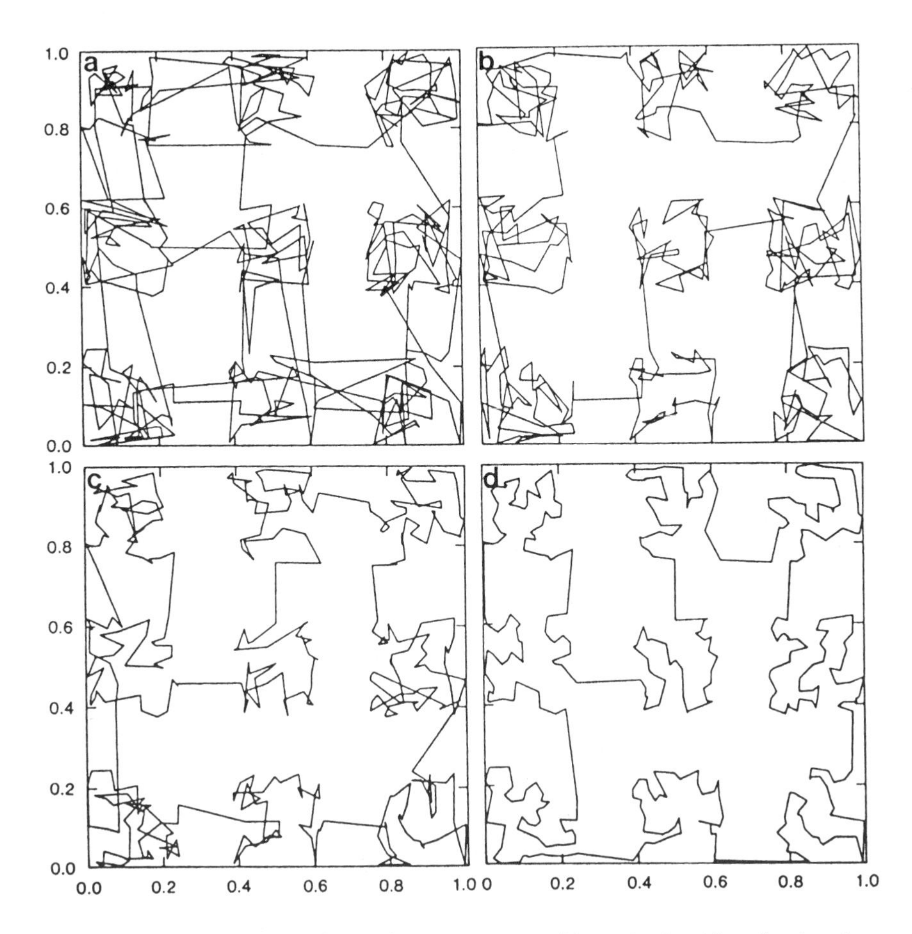

FIGURE 9.4 Tours of the traveling salesman to 400 cities, obtained by simulated annealing for four different annealing temperatures. The length of a tour is given by 400α. For a, $T = 1.2$ and $\alpha = 2.0567$; b, $T = 0.8$ and $\alpha = 1.515$; c, $T = 0.4$ and $\alpha = 1.055$; d, $T = 0.8$ and $\alpha = 0.7839$. (From S. Kirkpatrick, C. Gelatt, and M. Vecchi, "Optimization by Simulated Annealing," *Science*, vol. 220 (1983): 671–680, copyright 1983 by A.A.A.S.)

configuration at the given temperature, and α is proportional to the length of the tour. Note that the cities are distributed in clumps; the passage from one clump to another is stabilized at very high temperatures, while the finer details of the tour within the clumps only stabilize at lower temperatures.

9-1-2 The Partition Problem

The design of integrated circuits and printed circuit boards poses numerous combinatorial optimization problems for engineers. The partition of logic functions between integrated circuits, the placement of integrated circuits on boards, and the routing of connections are three examples among many. The partition problem provides a good illustration of the procedure. In the first phase of the design of a computer, we begin by drawing the schematics of the different logic functions needed (logic gates, flip-flops, counters, shift registers...) and the connections between them. The next step is to decide which VLSI circuits will contain these functions. With current technology, it is not possible to group all of these functions into a single integrated circuit, as the probability of obtaining a nonfunctioning circuit would be too high. Two contradictory and often frustrating demands must be taken into account:

- the circuits cannot contain too many logic functions, in order to lower the risk of failure; and
- interconnections between circuits should be kept to a minimum, since they are costly in terms of "pins" and of printed circuitry.

Therefore the logic functions must be distributed among several circuits in a way which best satisfies these two constraints. Let's take the case of the partition of functions, denoted i, between two integrated circuits A and B. If the function is implemented on circuit A, the corresponding variable S_i takes the value $+1$, or -1 if it is implemented on circuit B. The matrix of connectivities a_{ij} represents the necessary connections between the different logic functions. Its elements are equal to 1 if the connection is necessary, and 0 otherwise.

Assume that the connections implemented on the integrated circuit cost nothing, and only the connections which connect the two circuits are costly. The number of these connections, which is an important element in the cost function, is given by:

$$\sum\sum\nolimits_{i>j} \frac{a_{ij}(S_i - S_j)^2}{4} . \tag{9.1}$$

We must also take into account the relative densities of the circuits by the following term which is proportional to the square of the difference between the populations of both circuits:

$$\lambda \left(\sum_i S_i \right)^2 . \tag{9.2}$$

Generally, we choose λ to be of the order of $k/2$, where k is the average connectivity of a function.

Minimizing the sum of these two terms is equivalent to minimizing

$$\sum\sum\nolimits_{i>j} \left(\lambda - \frac{a_{ij}}{2} \right) S_i S_j . \tag{9.3}$$

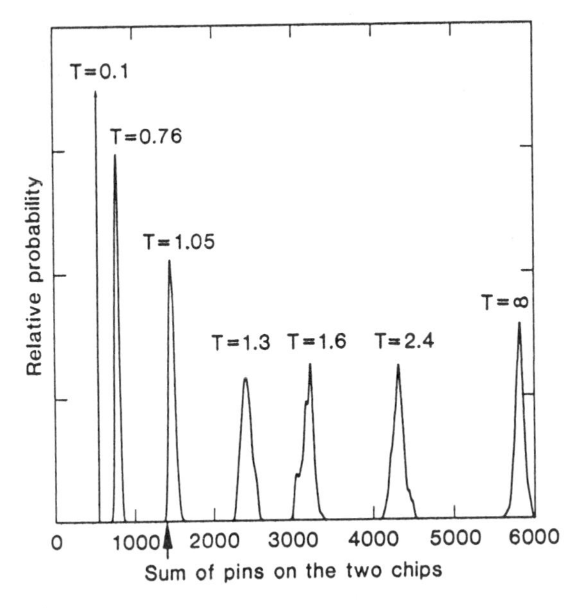

FIGURE 9.5 Histogram of the total number of pins of the two integrated circuits, obtained for different annealing temperatures. The arrow corresponds to the result obtained for quenching. (From S. Kirkpatrick, C. Gelatt, and M. Vecchi, "Optimization by Simulated Annealing," *ibid.*)

There is a striking similarity between this cost function and the energy function of a spin system. If there is no connection between i and j, the interaction term is positive, which then favors the implementation of the two functions in distinct circuits. Otherwise, the term is negative, which favors their implementation on the same circuit.

The elementary modification of the partition obviously consists of moving a function to a different circuit by changing the sign of S_i. The dynamics of simulated annealing is therefore formally equivalent to lowering the temperature of a spin glass. Figure 9.5 shows the results of a numerical simulation in which approximately 5000 logic gates are distributed between two integrated circuits. The histogram of the solutions is given in terms of the total number of output pins allowing the connections between the two circuits, plus the external connections (200 logic signals entering or leaving the system). We can see that the quality of the results improves considerably with annealing.

In summary, the simulated annealing technique is a very general method which can be used to find approximate solutions to combinatorial optimization problems. In fact, most cases do not require that we find the absolute minimum, and we can settle for an approximate value as long as the difference between the two energies is small. The appeal of this method lies in its broad generality and the fact that this procedure is easily parallelizable. For a given application, we need only determine the elementary modifications of the configurations which will be used for the trials. The computation of the corresponding energy changes must be very simple: at each trial it must only involve a small fraction of the data of the problem. The main difficulty in this method rests in the choice of the temperature plateaus. It is not always possible to define a protocol in advance. Most frequently, one proceeds in an empirical fashion: starting from high temperatures, one watches the system evolve, identifying the temperatures where the system no longer evolves; one anneals for a long time at slightly higher temperatures before continuing the descent.

There are numerous technological applications which require combinatorial optimization. Earlier we gave the example of computer-aided design, but the simulated annealing technique has already been used in domains as varied as the programming of garbage pickup in the city of Grenoble, or finding the optimal way to cut out the pieces of a suit on a roll of fabric. Many signal-processing applications can be approached from this point of view. We will discuss image processing in the section that follows.

9-2 Image Processing

Image processing is a set of techniques which improves images and enables them to be interpreted. The first step is the digitization of the image: it is decomposed into elementary cells called pixels. The image then becomes a matrix whose elements are the light intensities at each pixel. The intensities can be discrete variables, encoded with a certain number of bits: 1 (for black and white), 8, 12, 16, or real variables, in the ideal case. In the case of color images, the state of each pixel is described by three variables, one for each basic color. In the case of black and white images, a single intensity, the gray level, is sufficient.

We then apply to the digitized image a series of hierarchical transformations, from the lowest level—where the image is directly captured by the detectors (such as a photographic plate, a television camera, or a bank of particle detectors)—to the highest level—the interpretation of the image in terms of familiar objects. Recognition by the convergence toward an attractor, which we have mentioned in chapters 5 and 7, or by passing a pattern through a layered network (chapter 8), are high-level processing which allow the interpretation of the image. In this chapter we will discuss lower-level processing to reconstruct the image or to extract characteristic features from it.

9-2-1 Definitions and Cellular Methods

IMAGE RESTORATION An image acquired by a physical system is often 1) distorted and 2) noisy.

1. Distortions can come from:

- the optics (in the broad sense) of the system, aberrations, and diffraction; or
- from limits in the acuity of the detectors, such as the grain size of the photographic plate, the channels placed in front of the particle detectors, or the electronics of the television camera.

If the match between object and image was perfect, we could separate the surface of the object into elements which would each send a signal only to a single pixel. Because of the imperfections, each pixel also receives parasitic signals from the neighboring elements.

2. The noise can have an external, such as electronic, origin, but it can also be related to the physical nature of the detection procedure. In X-ray or γ-ray tomography, the cost of the instruments or the low intensity of the sources mean that the detectors are often used at the limits of their sensitivities, in a quantum regime. They are therefore particle counters. Being small, the particle flux fluctuates, and these fluctuations give rise to a high level of noise.

The aim of techniques to restore the image is to reconstitute an image which is as close as possible to the image that would be given by an ideal system, with no distortions or noise.

THE EXTRACTION OF CHARACTERISTIC FEATURES For the sake of efficiency, the image interpretation techniques are rarely applied at the level of the pixels of an image, even after restoration. The processing goes through intermediate steps during which certain characteristic features are recognized. These steps can involve the detection of contours, segmentation into areas with homogeneous intensities or textures, or specific techniques. In character recognition, for example, we can use skeletonization techniques, whereby the thick lines are replaced by thin ones, and the intersections of these lines with a predetermined set of straight lines are counted.

CELLULAR PROCESSING BY CONVOLUTION The cellular nature of the digitized image lends itself to cellular processing. In fact the first processing steps, at the lowest level, are of this type.

One simple way of reducing the noise consists of averaging the gray levels of neighboring pixels. We therefore perform a convolution between the image received and an operator representing the sum over the neighborhood v of each pixel:

$$x_1(i) = \sum_{j \in v} x_0(j) \,. \tag{9.4}$$

In this expression, the subscript 0 refers to the unprocessed image, and 1 to the result of the processing. Another use of the convolution operation is the detection of contours using the Laplacian operator:

$$x_1(i) = \sum_{j \in v} x_0(j) - z x_0(i) \tag{9.5}$$

where z is the number of pixels in the neighborhood. In regions where the intensity does not vary much, or varies in a regular way, this operator yields small values of

x_1. On the other hand, near a contour separating two parts of the image where the intensity changes markedly, it yields large values of x_1.

Several simple operations in image processing make use of convolutions with local operators. We can compare them to a single iteration on a network of cellular automata. Like other types of cellular processing, these operations are simple, can be carried out in parallel, and require no prior knowledge of the subject of the image.

Probabilistic Processing

A PRIORI DISTRIBUTION Another approach, of a very different nature, consists of using knowledge about the statistical properties of images to reduce the noise or to interpret the images. We expect *a priori* that the intensity will not vary abruptly from one pixel to the next, except along continuous lines that delimit the contours of objects. "Good" images are likely to be "regular": it is likely that any irregularities can be assumed to be defects due to noise or to certain imperfections in the equipment.

The idea is to translate our expectation of "regular" images into observation probabilities. Let the set of all of the images observable on an $N \times N$ matrix of pixels be:

$$\mathbf{X}^p = \{X^p{}_i\}$$

where p refers to the image and i to the pixel. To each image X^p of the set we can associate the following probability:

$$P(X^p = x) = \frac{\exp -(U(x)/T)}{Z} \tag{9.6}$$

where $U(x)$ is the "energy" associated with the image X^p, and T is the "temperature." Z, the normalization factor, is the sum of the exponentials associated with each of the other configurations. This formulation is reminiscent of statistical mechanics. All of the statistical information representing our knowledge about "valid" images is contained in the energy. The higher the energy of a configuration, the smaller the probability that it will be observed. We will give several possible expressions for this function below. An important property of the energy is that it behaves like a local variable, to the extent that the probability that pixel i has the value x_i depends only on the values of the neighboring pixels. Of course the extent of this neighborhood can vary. For example, we can account for the regularity of the figure in terms of the tendency of neighboring pixels to be at the same intensity by choosing an energy of the form:

$$U(\mathbf{x}) = \sum_i \sum_j f(x_i - x_j) \tag{9.7}$$

where f is an even function which decreases with the amplitude of the difference between the intensities of the neighboring pixels i and j. By analogy with the physics described in chapter 8, we are tempted to say that the neighboring pixels interact as a function of their intensity difference.

We also expect that the characteristic traits themselves will not be arbitrarily distributed. It is therefore logical to attribute probabilities of occurrence to them. Let's limit ourselves to the case of contours (Fig. 9.6). A *contour* delimits two regions of very different gray levels.

To each of the $Nz12$ pairs of neighboring pixels (i, j), we associate a binary variable $x^c{}_{ij}$ which describes the presence ($x^c{}_{ij} = 1$) or absence ($x^c{}_{ij} = 0$) of a contour. An image is now described by the set of pixels X^p and the set of contours X^c. The probability of the set of pixels plus contours is proportional to the exponential of an energy function given by a sum of terms related to the pixels and the contours:

$$U(\mathbf{X}^p, \mathbf{X}^c) = \sum_{i,j,i',j'} f(x^p{}_i - x^p{}_j) - [(a \cdot g\,(x^p{}_i - x^p{}_j) - b) \cdot (1 - x^c{}_{i,j})] - h\,(x^c{}_{i,j}, x^c_{i',j'}) \tag{9.8}$$

The first term, due to the interactions between the pixels, has been defined earlier.

The second term represents the interaction between neighboring pixels and the contour variable. If the intensity difference is large, it favors the state $x^c = 1$. Otherwise, state $x^c = 0$ corresponds to the lowest energy.

The third term corresponds to the interactions between the variables of neighboring contours. Continuous contours are favored compared to T intersections or crossings, which are *a priori* not very probable (Fig. 9.7).

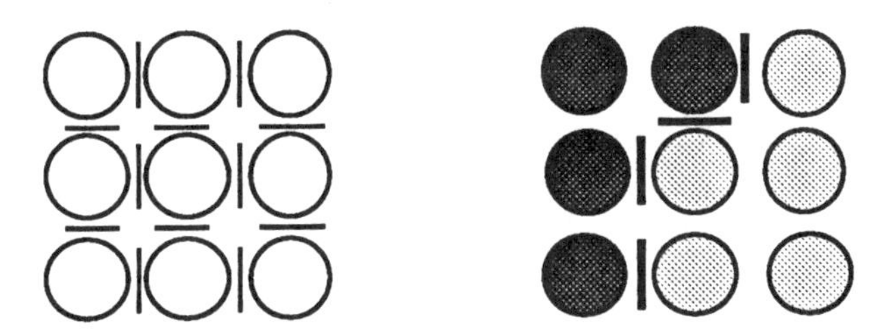

FIGURE 9.6 At the left, the pixels are represented by circles, and the contour elements by line segments between the pixels. A given configuration is shown on the right, with two regions of different gray levels. The thick line segments represent contour elements in state 1; those in state 0 are not represented.

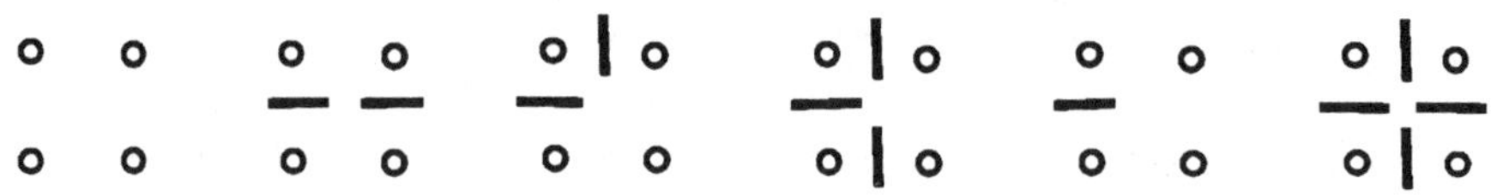

FIGURE 9.7 Interactions between contour elements. The energy includes a term which takes into account the relative arrangement of the contour elements. The six configurations shown above represent all of the possible configurations, to within a rotation. They are organized from left to right in order of increasing energy (and consequently of decreasing probability).

A POSTERIORI DISTRIBUTION In image processing we are concerned with the *a posteriori* distribution, i.e., the conditional probability of the corrected images $\{X^{p_i}\}$, given that the distorted image $\{Y_i\}$ was observed.

$$P_y(x) = P(X^p = x | Y = y) \,. \tag{9.9}$$

In other words, we must combine the information about the statistical properties of images, contained in the choice of the energy function, with the information obtained from the observed image. The simplest method to obtain the *a posteriori* distribution consists of adding to the energy a term proportional to the Hamming distance between the two images:

$$U_y(x) = U(x) + \sum_i \left(x_i^P - y_i\right)^2 \,. \tag{9.10}$$

The *a posteriori* distribution is then given by:

$$P_y(X^p = x) = \frac{\exp -(U_y(x)/T)}{Z} \,. \tag{9.11}$$

THE USE OF SIMULATED ANNEALING In fact, the image restoration problem is to propose, based on the observed image, the most probable restored image, in other words, given the above expression, the one that minimizes the energy $U_y(x)$. This then requires a simulated annealing technique. We start from an initial configuration which is the observed image, without contours, and we modify the pixel and contour variables by successive iterations using a Monte Carlo rule. The energy variations are calculated using the expression for $U_y(x)$. The temperature decreases steadily until the restored image is satisfactory. Of course, the most sensitive aspect of this method is the choice of the different parameters which define the energy function,

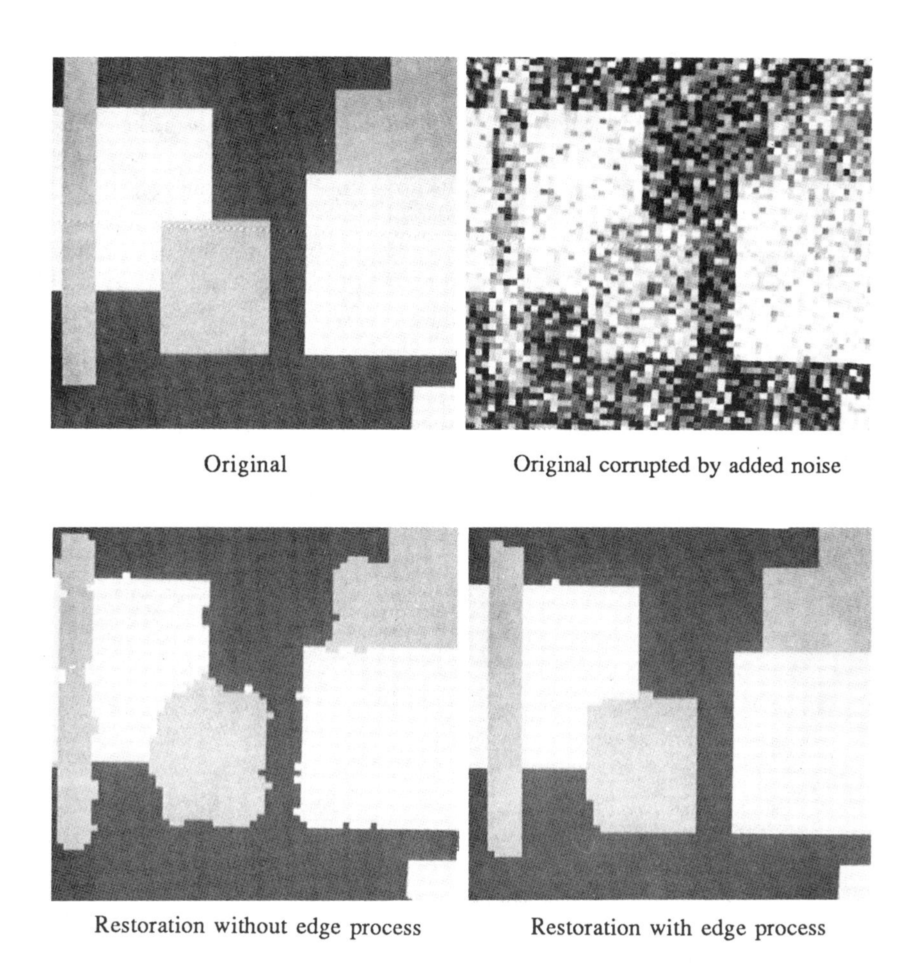

FIGURE 9.8 Image restoration and automatic contour tracing by simulated annealing (from S. Geman and D. Geman, "Bayesian image analysis," in *Disordered Systems and Biological Organization*, edited by E. Bienenstock, F. Fogelman-Soulié, and G. Weisbuch, NATO ASI Proceedings, vol. F20 (Springer, 1986). (a) Original figure; (b) noisy figure; (c) figure restored without use of contours; and (d) figure restored with contours.

as well as the profile of temperature decrease. Figure 9.8, taken from a paper by S. Geman and D. Geman, shows the results of this method in a very simple case.

The use of simulated annealing is a very promising approach for image processing of tomography or satellite images. The current difficulties involve the automation of the choice of the numerous parameters which define the energy function, and of course the volume of calculations, which could be carried out much more quickly by parallel processors.

The connection between magnetic systems at a finite temperature, probabilistic automata, and simulated annealing is particularly clear in the examples of the distribution of logic functions between different integrated circuits, and of image restoration. In both cases, the energy appears in the form of a sum of interactions between the components, and the updating procedure for the variables is random sequential iteration of probabilistic automata. This connection is less clear in other combinatorial optimization problems, such as the traveling salesman problem; in this case it amounts to the use of simulated annealing. Each of these three domains has benefitted from contributions by the other two:

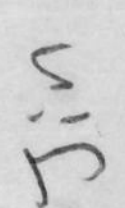

- The mathematical methods of statistical mechanics, the mean field and replica methods, come from magnetic systems.
- The Monte Carlo numerical methods are in fact random iterations of probabilistic automata.
- Finally, certain classical algorithms of combinatorial optimization have been used in the search for the minimum energy states of spin glasses (such as the algorithm of the Chinese postman).

References

The original paper by S. Kirkpatrick, C. Gelatt, and M. Vecchi, "Optimization by Simulated Annealing," *Science*, vol. 220 (1983): 671–680, is very readable.

A good reference on the traveling salesman problem is the collective book: *The Traveling Salesman Problem*, edited by E. Lawler, J. Lenstra, A. Rinnooy Kan, and D. Shmoys (John Wiley and Sons, 1985).

The classical reference on image processing is *Pattern Classification and Scene Analysis*, by R. Duda and P. Hart (Wiley-Interscience, 1973). Section 9.2.2 on probabilistic image processing is taken directly from the paper by S. Geman and D. Geman, "Bayesian image analysis," in *Disordered Systems and Biological Organization*, edited by E. Bienenstock, F. Fogelman-Soulié, and G. Weisbuch, *NATO ASI Proceedings*, vol. F20 (Springer, 1986).

The book by M. Mézard, G. Parisi, and M. Virasoro, *Spin Glass Theory and Beyond* (World Scientific, 1987), does a good job of illustrating the connections between the three fields: thermodynamics, optimization, and neural networks.

TEN
Random Boolean Networks

10–1 In Search of Generic Properties

In the preceding chapters, we have essentially taken an engineer's approach: our aim was to come up with recipes for constructing networks exhibiting a given behavior, or to search for the networks best suited to carry out a particular task. We can approach the network dynamics problem from a more general perspective by trying to characterize the properties of an "arbitrary," undedicated network—a network which was not constructed with the goal of accomplishing a specific task. This raises two preliminary questions:

- What is an arbitrary network?
- Which properties should be studied?

The notion of generic properties which we will now study is a possible answer to both of these questions. In fact, we are interested not in the particularities of a specific network, but in the orders of magnitude which we expect to observe in studying a set of networks with fixed construction principles. We therefore consider a set containing a large but finite number of networks. We choose some of these networks at random, construct them, and measure their dynamical properties. We then take the average of these properties, and we examine those which are fairly evenly distributed over the set of networks. An example will help to clarify these ideas.

Consider the boolean networks with connectivity $k = 2$, with a random connection structure. The dynamical variable we are interested in is the period for the set of all initial conditions and networks. Of course, this period varies from one network to the next. We have measured it for 10 randomly chosen initial conditions for 1000 different networks of 256 randomly connected automata, whose state change functions were generated at random at each node of the network. Figure 10.1 shows the histogram of the measured periods. This histogram reveals that *the order of magnitude of the period is ten* (this is the generic property), even though the distribution of the periods is quite large.

We can certainly construct special "extreme" networks for which the period cannot be observed before a million iterations. For this, we need only take networks which contain a random mixture of exclusive OR and EQUivalence functions (EQU is the complementary function of XOR; its output is 1 only if its two inputs are equal). But these extreme cases are observed only for a tiny fraction ($1/7^{256}$) of the set under consideration. We consider them to be pathological cases, i.e., not representative of the set being studied.

We then call *generic properties* of a set of networks those properties which are independent of the detailed structure of the network—they are characteristic of almost all of the networks of the set. This notion then applies to randomly constructed networks. The generic properties might be shown not to hold for a few pathological cases which represent a proportion of the set which quickly approaches 0 as the size of the network is increased. In general the generic properties are either:

- qualitative properties with probabilities of being true (or false, as the case may be) that are close to 1; or
- semi-qualitative properties, such as the scaling laws which relate the dynamical properties to the number of automata.

This chapter is devoted to the study of the generic properties of random boolean networks. It will contain a large number of examples of generic properties.

The notion of generic properties characteristic of randomly constructed networks is the basis for the theoretical biological models which we will describe in chapter 11. It has been extensively developed by physicists of disordered systems for the study of random microscopic systems such as glasses or macroscopic multiphase systems. They discovered (or rediscovered) many new concepts during the '70's. The notion of generic properties is similar to the notion of universality classes, developed for phase transitions. Without going into too much detail, we can say that the physical variables involved in phase transitions obey scaling laws which can be independent of the transition under consideration (such as transitions in magnetism, superconductivity, or physical chemistry) and of the details of the mathematical model which was chosen. These laws only depend on the physical dimension of the space in which the transition takes place (for us, this is three-dimensional space) and on the dimension of the order parameter. The set of phase transitions (and their mathematical models) which obey the same scaling laws constitutes a universality class. Moreover, this notion of a universality class has been expanded to include the regime transitions of continuous dynamical systems.

In fact, the first attempt to model a biological system by a disordered network of automata by S. Kauffman, a theoretical biologist, predates the interest of physicists in this subject. It is also based on the idea that the properties of disordered systems are representative of the vast majority of systems defined by a common average structure.

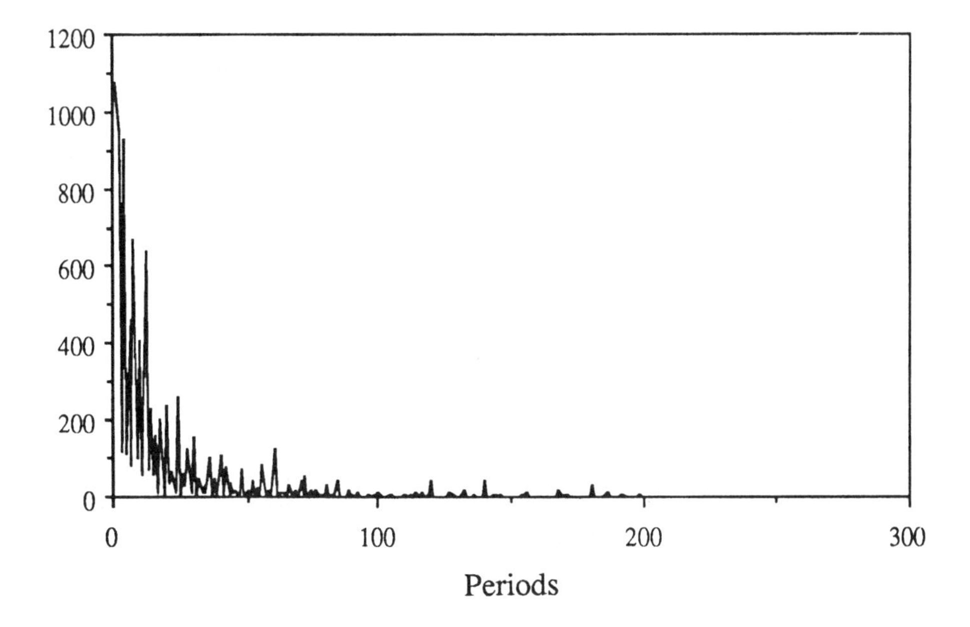

FIGURE 10.1 Histogram of the periods for 10 initial conditions of 1000 random boolean networks of 256 automata.

10–2 Scaling Laws for Periods

Stuart Kauffman studied the dynamical properties of random boolean networks as early as the late '60's. He constructed networks of boolean automata with an arbitrary input connectivity k. The state change function of each automaton is chosen randomly among the 2^{2^k} possible functions. The connections between the automata are also chosen randomly. The network is said to have random connectivity, in contrast to cellular networks, for example.

We are primarily interested in the statistical properties of these networks, which, in addition to being large, have large state spaces. Consequently, determining the complete iteration graph is out of the question. In order to study the attractors of a given configuration, we randomly pick a large number of initial configurations. For each of these, we let the network evolve toward an attractor. We save certain configurations, said to be references, at certain times, determined by doubling the previous sampling time (i.e., 2, 4, 8, 16, 32, etc.), and as soon as a reference is identical to a configuration saved at a previous time, we consider that the limit cycle has been reached. The period of the limit cycle is obtained by taking the shortest length of time it takes for a reference configuration to be visited twice. This reference configuration is then the only one to be saved. During the search for the limit cycle, we can determine if the limit cycles are distinct by comparing the current configuration to the set of reference configurations already obtained.

The results obtained by Kauffman show two distinct dynamical regimes, depending on the connectivity.

For networks of connectivity 2, the average period is proportional to the square root of N, the number of automata. The same is true of the number of attractors (see figure 11.1). In other words, among the 2^N configurations which are *a priori* possible for the network, the dynamics selects only a small number of the order of N which are really accessible to the system after the transient period. This selection can be interpreted to be an *organization* property of the network.

As the connectivity is increased, the period increases much faster with the number of automata; as soon as the connectivity reaches 4, the period as well as the number of attractors become exponential in the number of automata. These periods, which are very large as soon as the number of automata is greater than one hundred, are no longer observable, and are reminiscent of the chaotic behavior of continuous aperiodic systems. In contrast with the organized regime, the space of accessible states remains large, even in the limit of long times. We will see later that other dynamical properties of these discrete systems resemble those of continuous chaotic systems, and so we will refer to the behavior characterized by long periods as *chaotic*.

The scaling laws which relate the period and the number of attractors to the number of automata are generic properties of random boolean networks. They only depend on the connectivity of the network. It is certainly possible to construct pathological counter-examples: if we construct a network with connectivity 2 by using only XOR and EQUivalence functions, the period is exponential in the number of automata. But the set of networks constructed in this manner is but a tiny fraction of the set of random boolean networks. Consequently, this property is indeed generic.

The difference in the scaling laws as a function of the connectivity is indicative of very different dynamical behaviors in both cases. In this chapter we will introduce a series of tests, corresponding to generic properties, which will enable us to elucidate the differences between these two dynamical regimes.

The Case of Complete Connectivity

Despite twenty years of fruitless labor, Kauffman's results have still not been proven. Only two extreme cases, $k = 1$ and $k = N$, are well understood.

- For $k = 1$, the networks are composed of independent "crabs" whose dynamics has been explored in chapter 2 (2.2.3).
- For $k = N$, the case of *complete connectivity*, the state of each automaton depends on all of the others, including itself. In order to specify the state change rules, it is necessary and sufficient to write the complete table of the $M = 2^N$ successors of each of the configurations of the system. We can then interpret each configuration as the binary representation of a whole number

between 0 and $M-1$. The stated problem is then equivalent to the problem of mapping a finite set of M integers onto itself.

For a random network, the successor table, like the iteration graph, is completely random: the probability that an arbitrary configuration will succeed another is $1/M$.

Consider the iteration subgraph:

$$C_0 \longrightarrow C_1 \longrightarrow C_2 - \cdots \longrightarrow C_i - \cdots \longrightarrow C_{k-1}$$

which starts at the configuration C_0 and reaches the configuration C_{k-1} after $k-1$ iterations, such that all of the configurations $C_0, \cdots C_i, \cdots C_{k-1}$ are different. $P(k)$, the probability of this subgraph, is given by the product of the probabilities that each of the i configurations is different from the $i-1$ preceding ones:

$$P(k) = (1-1/M)\cdot(1-2/M)\cdot(1-3/M)\cdots(1-(k-1)/M)\,. \tag{10.1}$$

$P(k)$, the product of terms less than 1, is also less than 1; for small values of k, $P(k)$ is close to 1. In fact, if we take its logarithm, we see that we can approximate $P(k)$ by a gaussian of the form:

$$P(k) = \exp -\frac{k(k-1)}{2M}\,. \tag{10.2}$$

This expression holds for $1 \ll k \ll M$.

It follows that $P(k)/M$ is the probability that an arbitrary configuration is the C_k configuration which follows C_{k-1}, whether or not it belongs to the subgraph. In particular, this is the case for C_0. $P(k)/M$ is therefore the probability that C_0 belongs to a cycle of length k. $n(k)$, the average number of cycles of length k, is therefore:

$$n(k) = \frac{M \times P(k)}{M \times k}\,. \tag{10.3}$$

In other words, $n(k)$ is the average number of points which belong to a cycle of length k, divided by the length of the cycle. The distribution of cycles is strongly shifted toward shorter cycles. On the average there exists one cycle of length 1.

The total number of cycles corresponds to the sum of the $n(k)$ for k going from 1 to M. We can show that this sum tends toward $1/2 \log(M)$, or $N/2 \log 2$. (We can approximate this limit by noticing that the width of the gaussian is $\sqrt{M}$. The order of magnitude of the sum is thus obtained by summing the harmonic series in $1/k$ until $k = \sqrt{M}$.)

The first scaling law can be stated as follows: the number of attractors varies linearly with the number of automata.

From an arbitrary configuration C_0, the probability that the loop closes at C_i, giving a period of length $k-i$, is still $P(k)/M$. In particular, this probability is independent of i. We therefore obtain the average length of a cycle by a double sum over i and k of the period $k-i$ weighted by the probability $P(k)$. By approximating the sums by integrals and $P(k)$ by the gaussian, we obtain:

$$T = \frac{1}{M} \int_0^{M-1} dk \int_0^{k-1} (k-i) \exp\left[-\frac{k(k-1)}{M}\right] dk\,. \tag{10.4}$$

In this way we obtain a term proportional to $\sqrt{M}$.

The average period is therefore an exponential function of the number of automata:

$$T = \alpha 2^{N/2}\,. \tag{10.5}$$

This is the second scaling law. α is 0.63.

Finally, both of these results taken together indicate that the attractors occupy a large fraction of the configuration space.

Unfortunately there exists no exact analytic solution for the intermediate cases, for connectivities between 1 and N.

10–3 Cellular Networks with Random Functions

The study of inhomogeneous networks with cellular connectivity enables us to readily visualize the states of the automata during the course of the iterations, and thus to better understand the dynamics of random networks. Two researchers at M.I.T., H. Hartman and G. Vichniac, proposed that structures with periodic connections, where the functions of the automata are different at each node be called INCA (for INhomogeneous Cellular Automata).

One-dimensional INCAs can consist of a linear chain with interactions between nearest neighbors. One very simple structure is the chain of automata with 2 inputs which we discussed in chapter 3 (3.3).

Figure 10.2 shows the evolution of a linear INCA composed of a random mixture of 30% OR functions and 70% XOR functions. In this study, as well as in the studies that follow of two-dimensional networks, the boundary conditions are periodic: the automaton at the extreme right of the chain is connected to the automaton of the extreme left. The first row indicates the function of each automaton: O represents the OR function, and X represents the XOR function. The following rows represent the state of the network at each time interval, with time increasing toward the bottom of the figure. Asterisks * represent automata in state 1, and periods . represent automata in state 0. Recall (see chapter 3) that a linear network of two input automata can be divided into two independent subnetworks when the number of automata is even. Figure 10.2 therefore only represents one of these subnetworks.

The shaded rectangles surround subnetworks of nodes which are invariant during the course of the iteration, after a transient period which varies according to the subnetwork. These shaded walls isolate oscillating areas. The unshaded areas

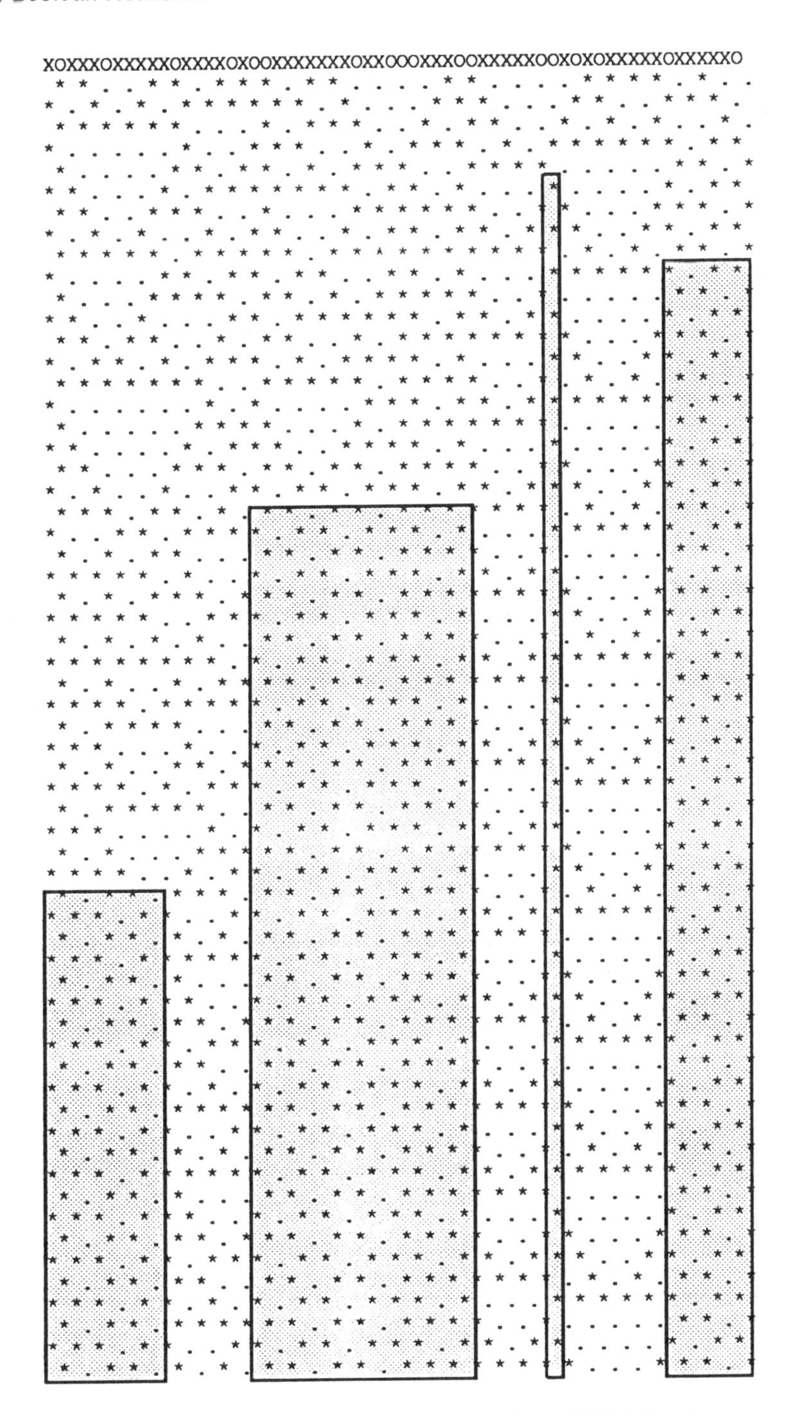

FIGURE 10.2 Linear INCA composed of a mixture of OR and XOR functions.

oscillate independently from each other, as the stable walls block all transmission of the signal between oscillating areas. On this figure we can measure the periods of the oscillating subnetworks: from left to right, they are 10, 4, and 6. The period of the network is their smallest common multiple, which is 60.

The origin of stable walls can easily be understood from this example. As soon as a pair of neighboring OR functions occupies the 11 configuration, it stays in this configuration indefinitely. Indeed, in order for an OR function to stay in state 1, it is sufficient for one of its inputs to remain in this state. The OR function is therefore a *forcing function*, with a state which is fixed by a single one of its inputs. This state is the *forced state*, and the input which allows this operation is the *forcing input*. Both inputs of an OR function are forcing inputs.

A pair of neighboring OR functions is a *forcing structure*, since each forced state of a cell is the forcing state of the other. Forcing structures spread out from pairs of OR functions, as we can see on figure 10.2.

As we increase the size of the network, the sizes of the oscillating areas increase very slowly, as does their period. Consequently, the period of the network also goes up very slowly with the size.

In conclusion, the temporal organization of the network into short period cycles is linked to its spatial organization into oscillating subnetworks separated by stable walls, and the existence of stable walls is due to the forcing structures. This conclusion remains valid for any random mixture of boolean functions with two inputs.

10–3–1 Functional Structuring

The same sorts of observations can be made on a two-dimensional cellular structure such as the one shown in figure 10.3, for which the structure of the connections alternates from one site to its neighbor.

The state change functions are chosen randomly among the 14 non-constant boolean functions with two inputs (excluding the constant functions which are fixed in the 0 or 1 states). Figure 10.4 shows four configurations of the same network which belong to limit cycles obtained from different initial conditions.

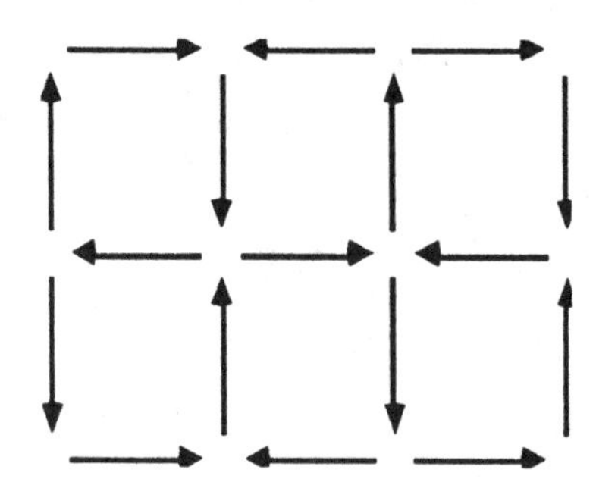

FIGURE 10.3 Cellular connection structure for automata with two inputs.

```
. 0 0 0 . . * . 0 1 * . 1 0 1 *     . 1 1 1 . . * . 0 1 * . 1 0 1 *
* 0 1 0 1 0 . . 0 1 * . 1 1 1 1     * 0 1 0 1 0 . . 1 1 * . 1 0 1 0
0 . . 0 0 0 0 . . 0 1 1 * 0 1 1     0 . . 1 1 0 1 . . 0 1 1 * 0 1 1
. . 0 0 1 0 0 * * 1 1 0 * . * .     . . 0 0 1 0 0 * * 0 1 1 * . * .
* * 0 1 * . 0 0 . 0 * * . * * .     * * 1 0 * . 0 0 . 0 * * . * * .
. 0 0 0 . . 0 1 . * 1 0 * . . *     . 0 0 0 . . 0 0 . * 1 1 * . . *
. . * 0 . * 0 0 1 . 1 1 1 . . .     . . * 0 . * 0 0 1 . 1 1 1 . . .
. 1 0 0 . 1 0 . * 0 0 * 1 . . .     . 0 0 1 . 0 0 . * 1 0 * * . . .
0 1 1 1 0 1 1 0 0 * * * 1 0 * 1     . 1 0 1 1 1 1 0 0 * * * * * * .
1 * 0 . 0 1 1 0 1 1 . 1 1 1 0 0     * * . . 0 1 1 1 1 0 . * . * * .
0 0 1 * 1 0 0 1 * . 1 0 . 1 1 1     . * * * 1 0 0 1 * . 1 1 . . * .
1 1 * * 0 1 0 . * * 0 0 * 1 1 0     . * * * 0 1 0 . * * 0 1 * * . .
* . * * 0 0 0 * . . . 0 0 1 0 .     * . * * 0 0 0 * . . . 1 0 1 1 .
* * * . * . . . * . * . 1 0 . *     * * * . * . . . * . * . 1 0 . *
. * . . . * . * * * . 1 0 1 . .     . * . . . * . * * * . 0 1 1 . .
* . . * * * * * * . . 1 1 . . .     * . . * * * * * * . . 1 1 . . .

. 0 1 0 . . * . 1 0 * . 1 1 1 *     . . * . . . * . 1 0 * . 0 1 0 *
* 0 0 0 0 0 . . 1 1 * . 0 1 0 1     * . * . * . . . 0 0 * . 0 1 1 1
0 . . 0 1 0 1 0 . 1 1 1 1 1 1 0     1 . . . * . * 1 . 0 1 1 1 1 0 0
. . 1 0 0 0 0 1 1 0 0 1 1 0 * .     . . . . * . . 1 0 1 0 0 0 0 * .
* * 1 1 * . 0 0 0 1 1 0 0 * * .     * * * * * . 0 0 0 0 1 0 0 * * .
. 0 1 0 . . 1 0 0 1 0 1 1 . . *     . . . . . . 1 1 1 1 1 0 0 . . *
. . * 0 . * 0 0 1 . 1 0 1 . . .     . . * . . * 0 1 1 . 1 0 1 . . .
. 1 0 0 . 0 1 . 1 1 1 * 1 . . .     . 0 1 1 . 1 1 . 0 0 0 * 1 . . .
0 0 1 0 0 1 1 1 0 * * * 0 1 * 0     0 0 1 0 0 0 1 1 0 * * * 0 1 * 0
1 * 0 . 1 1 0 1 0 0 . 1 0 1 1 0     1 * 0 . 1 0 0 0 1 1 . 1 0 1 1 0
1 1 0 * 1 1 0 0 * . 1 1 . 0 0 0     1 1 0 * 1 1 0 0 * . 1 1 . 0 0 0
0 1 * * 0 0 1 . * * 0 1 * 1 0 0     0 1 * * 1 1 0 . * * 0 1 * 1 0 0
* . * * 0 1 0 * . . . 1 0 1 0 .     * . * * 0 1 0 * . . . 1 0 1 0 .
* * * . * . . . * . * . 0 0 . *     * * * . * . . . * . * . 1 1 . *
. * . . . * . * * * . 0 0 1 . .     . * . . . * . * * * . 0 1 1 . .
* . . * * * * * * . . 0 0 . . .     * . . * * * * * * . . 1 1 . . .
```

FIGURE 10.4 Limit cycles of a 16 × 16 random boolean network. These configurations were obtained by starting from four different initial configurations. The 0's and 1's correspond to oscillating nodes, and the asterisks * and periods . correspond to nodes which, during the limit cycle, are fixed in the states 1 and 0, respectively.

In two dimensions, we also observe the spatial structuring characteristic of short period networks. In fact, the connection between short periods and the organization into functionally independent subnetworks is not a consequence of the cellular structure. It also exists for randomly connected networks of boolean automata with two inputs. In the absence of a two-dimensional representation, we can verify using a computer that during the limit cycle, the sets of oscillating nodes are unconnected.

31	533	533	533	0	0	0	0	999	999	0	0	999	999	999	31
31	533	533	533	533	533	0	0	999	999	0	0	905	999	999	999
999	0	0	533	533	564	564	749	0	999	936	936	717	936	999	999
0	0	564	564	533	564	564	749	749	999	936	936	655	655	0	0
0	0	564	564	0	0	999	999	749	999	718	655	655	0	0	0
0	564	564	564	0	0	999	999	749	749	936	936	655	0	0	0
0	0	0	564	0	0	999	999	999	62	936	936	936	0	0	0
0	999	999	999	0	999	999	62	780	936	936	62	407	0	0	0
438	999	999	999	999	999	999	999	999	62	62	62	438	438	0	438
438	0	438	0	999	999	999	999	999	999	62	438	438	438	438	438
438	438	438	0	999	999	999	999	63	63	782	782	0	438	438	438
438	438	0	0	999	999	999	0	63	63	782	782	0	438	438	438
31	155	155	0	999	999	999	0	63	63	0	782	782	937	937	31
31	155	155	155	0	0	0	0	63	63	0	0	999	999	0	31
31	0	0	0	0	0	0	0	0	0	0	875	999	999	31	31
0	0	0	0	0	0	0	0	0	0	0	875	875	31	31	31

FIGURE 10.5 Statistics on the number of initial conditions for which the nodes of the network oscillated (for a total of 999 initial conditions).

10–3–2 The Plateaus

The spatial structure depends on the initial conditions, as is shown in figure 10.4. Depending on these conditions, certain nodes either oscillate or remain fixed. We can analyze this dependence by evaluating the probability of oscillating for each of the nodes over the set of all initial conditions. Figure 10.5 was obtained for the same network as figure 10.4. For 999 randomly chosen initial conditions, we computed the number of times each node oscillated during the limit cycle.

On this figure, the 0's correspond to nodes which were stationary during all the limit cycle: their set constitutes the *stable kernel*. The 999's correspond to nodes which always oscillated. The intermediate values are grouped into plateaus corresponding to groups of neighboring nodes: 31, 62, 63, 155, 407, 533, 564, 655, 717, 749, 780, 782, 875, 905, 936, 937. These discontinuous plateaus are indicative of the number of different limit cycles and the sizes of their basins of attraction.

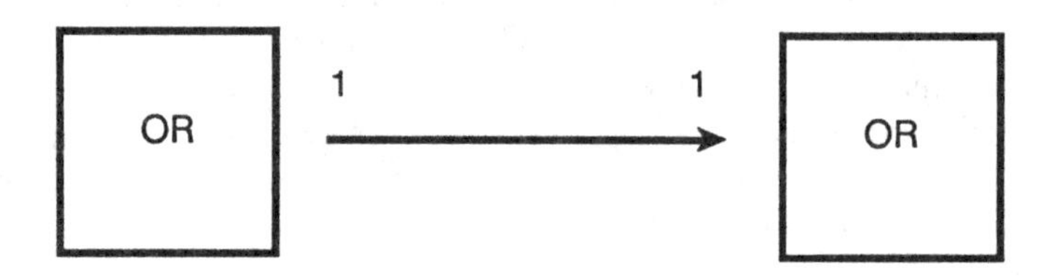

FIGURE 10.6 A forcing edge.

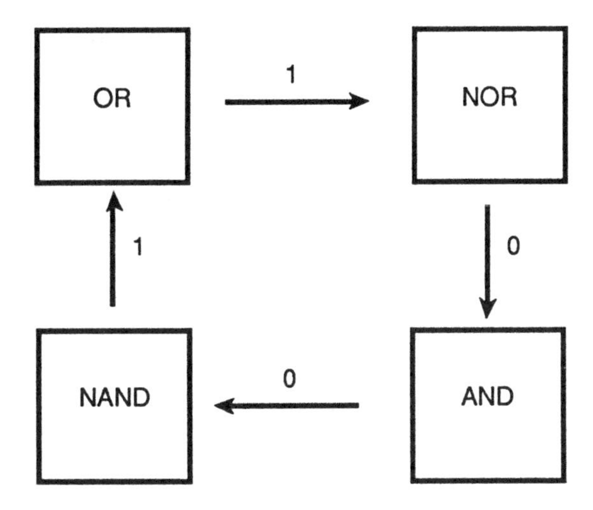

FIGURE 10.7 A forcing loop.

10–3–3 Forcing Structures

Forcing structures are responsible for the existence of the stable kernel. 12 out of the 14 boolean functions used are forcing functions, which remain fixed if one of their inputs is fixed at the forcing value.

The structure shown in figure 10.6 is a *forcing edge*: the forced output of the OR function on the left is also the forcing value of the OR function on the right.

Figure 10.7 represents a *forcing loop*, where each function forces the following one. All loops of four forcing functions are not necessarily forcing loops—the sequence of forcing and forced values must also occur in the proper order. These notions are generalizations of the concept of frustration introduced for a connectivity of 1 in chapter 2. Forcing loops are a generalization of the concept of unfrustrated loops; they are the seeds of the stable kernel. Other elements are recruited transiently, such as forcing functions with a forcing input connected to the forcing loop, or even non-forcing functions with both inputs determined by their connections with the forcing loop.

The existence of stable kernels, and consequently of short periods, depends on the relative proportion of forcing functions. For a connectivity of 1, all of the functions are forcing functions. We saw in section 2.2.3 that these networks were always composed of subnetworks, and that their periods were short. If the connectivity is 2, 14 of the 16 boolean functions are forcing functions: it follows that random networks have short periods and are structured. On the other hand, as soon as the connectivity reaches 4, the proportion of forcing functions falls to 4%, which explains the absence of forcing structures and of structuring, and the presence of long periods.

10–3–4 The Phase Transition

The connectivity parameter is an integer. It is interesting to introduce a continuous parameter in order to study the transition between the two regimes: the organized regime for short periods, and the chaotic regime corresponding to long periods. B. Derrida and D. Stauffer suggested the study of square networks of boolean automata with four inputs. Figure 10.8 shows the structure of the connections.

The continuous parameter p is the probability that the output of the automaton is 1 for a given input configuration. In other words, the networks are constructed as follows. We determine the truth table of each automaton by a random choice of outputs, with a probability p of the outputs being 1. If $p = 0$, all of the automata are invariant and all of the outputs are 0; if $p = 1$, all of the automata are invariant and all of the outputs are 1. Of course the interesting values of p are the intermediate values. Because of the symmetry between 0 and 1, the behavior of the network is symmetric around $p = 0.5$. It is therefore sufficient to study the behaviors on the $[0, 0.5]$ interval. If $p = 0.5$, the random process described above evenly distributes all of the boolean functions with four inputs; we therefore expect the chaotic behavior predicted by Kauffman. On the other hand, for values of p near zero, we expect a few automata to oscillate between attractive configurations composed mainly of 0's, corresponding to an organized behavior. Somewhere between these extreme behaviors, there must be a change of regimes. The critical value of p is 0.28. For smaller values, we observe small periods proportional to a power of the number of automata in the network. For $p > 0.28$, the period grows exponentially with the number of automata.

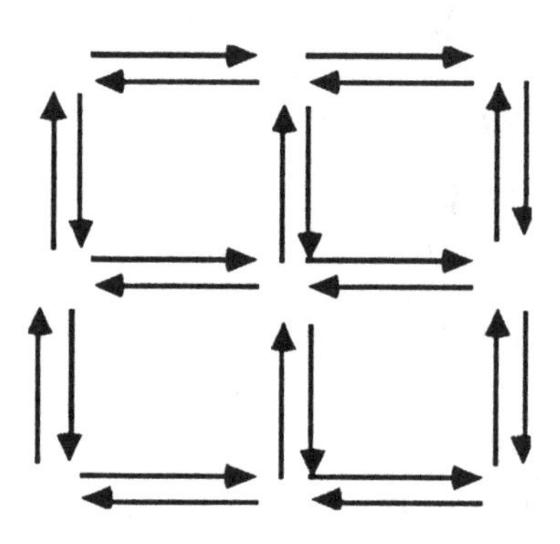

FIGURE 10.8 Connection structure for automata with four inputs.

```
 1  1  4  4  8  8  8  1  1  4  4  1  1  1  1  1
 1  1  1  1  8  8  4  1  1  1  1  1  1  4  1  1
 4  1  1  1  8 72  4  1  1  1  1  1  1  4  1  4
 4  1  1 72 72 36  4  1  2  1  1  1  1  1  1  1
 4  4  1 18  1 18 36  2  2  2  1  1  1  1  6  1
 4  1  1 18 18 18 36  1  1  1  1  1  1 12  6 12
12  1 18 18 18 18 18  1  1  1  1  1 12 12  1 12
 1  1 18 18 18  1 18  1  1  1  1  1  1 12 12 12
 1  1  1  1  1  1  1  1  1  4  4  1 12 12  1  1
 1  1  1  1  1  1  1  1  1  1  4  4  4 12  1  1
 1  4  4  1  1  1  1  1  1  1  4  4  1  4  1  4
 1  1  1  1  1  1  1  1  1  1  1  4  1  4  4  4
 1  1  1  1  2  2  1  1  1  1  1  1  1  1  4  1
 1  1  1  1  1  1  1  1  1  1  1  1  1  1  4  1
 1  1  1  8  8  1  1  1  1  1  1  4  4  1  1  1
 1  1  8  8  8  8  1  1  1  1  4  1  1  4  4  1
```

FIGURE 10.9 Local periods of a network in the organized regime. $p = 0.2$.

```
  1*********  1  1  1  1  1  1  1  1  1  6  6  1
  1******************  1  1  1  1***************
***************************  1  1  1  1******  1
******************  1  1******  1  1  1  1******  1
******************  1***************  1*********
***************  2  2  2  2***  4***  1*********
************  1  1  1  2  1  1******************
*********  1  1  1  1  1  1  1******************
*********  1  1  1  1  1  1*********************
***************  1  1  1******************  1***
************  1  1  1  1************************
***  1  1***************************  1*********
***  1  1***************************  1***  1  1
  1  1  1*************************** 12*** 12  1
  1  1************************  1  1 12  6  6  1
  1  1******  1  1******  1  1  1  1  1  2  6  1
```

FIGURE 10.10 Local periods in the chaotic regime. $p = 0.3$. Groups of three asterisks *** correspond to nodes which have periods too long to be measured.

The study of local periods is also interesting. Each automaton potentially oscillates with a certain period during the limit cycle. The period of the network is the least common multiple of all of these periods. Figures 10.9 and 10.10 show these periods.

We can see in figure 10.10, in the case of the chaotic regime, that certain nodes have long periods that cannot be determined for the duration of the simulation. The other nodes have relatively short periods.

10–3–5 Percolation

The notion of *percolation* is important in the study of generic properties of disordered cellular networks. We say that a region which is homogeneous (with respect to a certain property, which in figure 10.10 corresponds to the region of automata with long periods) percolates if we can go from one end of the network to another from within this region. On figure 10.10 the region of long periods percolates, isolating the regions of short periods. On figure 10.9 it is the region of fixed automata which percolates. If a given set percolates, the complementary set is no longer connected.

```
. 3 3 3 3 1 . 3 3 9 9 9 9 9 . . . . . . 7 7 7 . . . 9 . . .
. . . 3 3 3 3 3 3 3 1 . 9 . . . . . . . 7 7 7 7 . 9 9 7 . 2
2 . . . . 3 . 3 3 3 3 3 . . . . . . . 9 9 . . 7 7 7 6 9 . .
. . . . 3 3 3 . . . 8 3 8 8 . 6 6 . . . 9 2 2 . 7 7 8 9 . .
. . . . 3 9 . . . 3 8 8 8 6 . . 1 . . . . 2 2 2 7 . 8 . . .
. . . . . 9 . . . 3 . . 8 . 5 5 3 . . . . . . . . 9 . . . .
. . . . . . . . . . . . . . 5 5 4 . . . . . 4 4 . 9 9 9 9 .
. . . . . . . . . 2 . . 2 . . . 4 1 . 4 4 4 . 4 9 9 . . . .
. . . . . . 9 9 . 2 2 2 2 . . . . 4 4 4 4 4 9 9 9 9 . . . 9
. 9 9 . . 9 9 9 . 2 2 . . . . . . . 4 4 . . 9 4 . . . . . 9
9 9 9 . . . 9 . 2 2 . . . . . . . . 4 . . . 9 9 . . . . . .
9 9 9 9 . . 9 . . 2 . . 4 4 . . . . 4 . . . 9 9 . . . . . .
9 9 9 . 9 9 7 . . . . . 4 4 . . . . 4 9 . 9 9 . . . 9 . . 9
9 . . . . . 2 . . . . . . . . . 9 9 9 9 8 8 3 . . . 9 . . 9
9 . . . 1 2 2 . . . 9 9 . . 9 9 9 9 9 8 8 . 1 1 . 9 9 . 9 9
. . 1 . 1 . . . . . 9 . . . 9 9 . 9 9 8 . . . . . 9 9 . . 9
9 . 1 1 1 . . . 9 9 9 . . . 9 9 9 9 7 9 9 9 . . 9 9 9 . . 9
9 9 7 4 2 2 . . 9 . 9 9 . . . 9 9 9 9 9 9 9 . 9 9 . . . 1 1
. 9 9 9 . . . . 9 . 1 9 9 . . . . . 9 9 . 9 . 9 9 9 . . 5 .
. . 8 6 . . . . 9 . 1 9 . . . . . . 9 9 . 9 9 . 9 9 9 9 4 5
5 3 5 2 2 . . . . . . . . . . . . . . 9 . 9 9 9 . . . . 9 5
5 . . . . . . . . . . . . . . . . . . 9 . . . . . . . . 9 5
5 . . . . . . . . . . . 9 9 . . . . 9 9 9 . . . . . . . 9 9
. 9 9 . 9 9 . . . . . . 9 . . . 9 9 9 9 . . . . . . . . . .
. 9 9 9 9 . . . . . . . 9 9 . . . . 9 9 . . . . . . . . . 9
9 . 9 9 9 9 . . . . . . 9 . . . . . 9 9 . . . . . . . . . 9
9 . 9 9 9 . . . . . 9 . 9 9 . . . . 8 9 . . . . . . . 9 9 9
9 9 9 9 . . 9 . . . 9 8 9 8 . . . . . . . . . . . . . . 9 .
. . . 9 . . 9 . . . . 8 8 8 . . . 9 9 9 9 9 9 9 . . . 9 9 .
. . . 3 . 9 9 9 3 . . . 9 9 9 . . 9 . . 7 7 7 9 . . 9 9 9 .
```

FIGURE 10.11 Oscillation statistics for a random boolean network in the organized regime ($p = 0.21$). We observe the percolation of the stable kernel, represented by dots.

The percolation of the stable kernel is a good test of the organized or chaotic nature of the network dynamics. Figures 10.11 and 10.12, which correspond to two random networks with $p = 0.21$ for the first and $p = 0.28$ for the second, show the statistics on oscillating nodes for 9 initial conditions. The stable nodes are represented by dots. In figure 10.11 we observe the percolation of the stable kernel. The oscillating areas are therefore isolated. On the other hand, for $p = 0.28$, the set of oscillating nodes percolates from top to bottom (but not from left to right!). We also note that almost all of the oscillating nodes, with the exception of 8, belong to the same connected set. The statistical study of the percolation of the stable kernel as a function of p for increasingly large networks allows us to determine the percolation threshold, above which the stable kernel percolates in a network whose size can go to infinity: this is $p_c = 0.28 \pm 0.02$. Of course, because of symmetry, we also find an organized behavior for p larger than 0.72.

```
. . . . . . . . . . 9 9 9 9 2 9 9 9 8 6 6 6 1 9 9 7 . . . .
. . 3 3 3 . . 9 . . . . 9 9 9 9 9 9 6 5 6 6 6 . . . . . . 1
. . . . . . . 9 9 9 . . 9 9 9 9 9 9 7 6 6 . . . . 9 . . . 1
. . . 9 . 3 9 9 9 . 9 9 9 9 9 9 9 9 8 8 6 2 . . . 9 9 9 9 .
. . . 9 3 3 3 . . . 9 9 9 9 9 9 9 9 9 9 6 2 . . . 9 9 9 9 9
. . . 8 . . . . . . 9 9 9 9 9 9 8 9 9 9 8 9 9 9 9 9 9 8 8 9
. . . 1 1 1 1 1 1 9 9 9 9 9 9 8 9 8 9 8 5 8 9 9 5 9 9 9 9 9
1 . . . 1 1 1 1 1 8 8 9 9 9 9 9 9 9 9 9 9 9 6 9 9 9 . . 9 9
9 9 . . . . 3 3 3 2 9 9 9 3 3 6 9 7 9 9 9 5 2 9 9 9 . . 9 9
1 1 1 . . . 3 3 3 3 3 9 3 3 3 . . . . . 9 9 2 9 7 9 9 . . .
1 . 1 . . . 3 3 9 9 3 3 . 3 . . . . . . 9 9 . 9 9 . 9 9 . .
. . . . . . 3 9 9 8 3 3 3 3 2 8 . . 9 9 9 9 9 9 9 9 9 9 9 9
9 . 9 9 . . . 1 1 9 9 3 3 3 2 8 . . . 9 9 . . 9 9 9 9 9 9 9
9 9 9 9 . . . 1 . 8 9 8 3 . 8 8 . . . 9 . . . . . 9 9 9 . 9
9 9 9 9 . 9 9 9 9 9 9 8 8 8 8 1 1 . 5 . . . . . . . . . . .
9 9 9 9 . 9 9 9 9 9 9 8 8 8 . 1 1 . 5 . . 9 9 9 . . . . . .
9 7 9 9 9 . . . 8 9 2 8 9 8 . . . . 8 . . 3 8 . . 2 7 2 2 .
9 9 . . 9 9 . . 9 8 8 8 9 8 8 . . 9 9 9 9 3 9 . . 7 6 6 . 9
9 7 7 . . . . . 9 9 8 8 9 9 . . . 9 9 9 9 9 9 9 . . 6 6 9 9
9 9 . . . 9 9 9 9 8 9 8 8 9 . . . 9 9 9 9 9 9 8 . 9 6 6 . 4
. . . . 9 9 9 9 9 9 9 8 9 9 9 9 9 9 9 9 9 9 8 9 9 9 9 6 . 4
. . . . 9 9 8 9 9 7 8 9 9 9 9 7 . . 9 9 9 9 9 9 9 1 . 6 . .
. . 9 9 . 9 9 9 9 8 8 9 . . . . . . . 9 9 . . 9 . 1 3 . . .
. . 9 . . 7 7 9 . . 6 9 . . . . . . . 9 9 . . 9 . 2 3 3 3 .
. . . . 7 7 7 2 9 9 1 1 . . . . . . . . . . . 9 . 2 3 3 . .
. . . . 7 7 7 3 9 . 2 . . . . . . . . . . . . 1 . 3 9 3 3 .
. . . . . . . . . 2 2 2 . . 9 9 . . . 9 . . 1 1 1 3 9 . . 9
9 9 . . 3 . . . . . . 9 2 2 2 9 9 9 9 9 . 4 4 . 1 3 . . . .
. 9 9 . 3 . . . 9 9 9 9 9 9 . 9 9 9 9 1 1 1 4 1 9 9 . 9 9 .
. 9 9 . . . . 9 9 9 9 9 9 9 9 9 9 9 9 6 6 6 3 2 7 7 7 . . .
```

FIGURE 10.12 Oscillation statistics for a random boolean network in the chaotic regime ($p = 0.28$). We observe the percolation of the set of oscillating nodes.

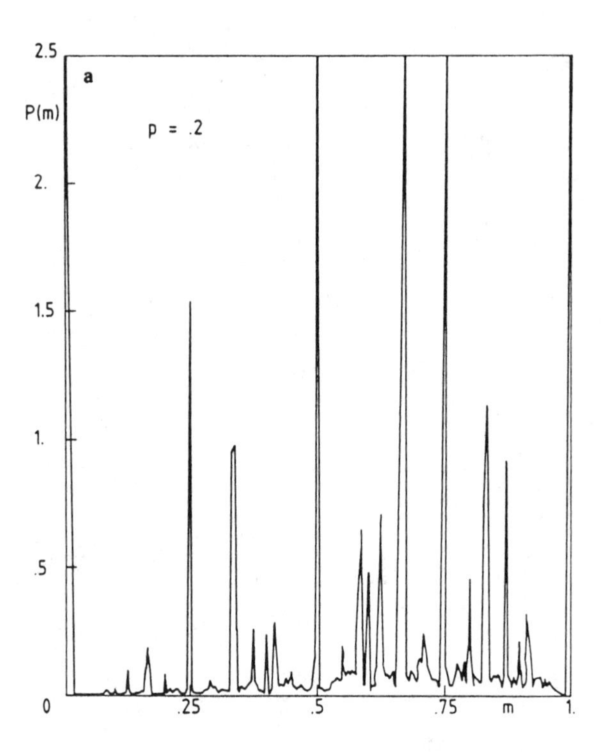

FIGURE 10.13 Histogram of the magnetizations in the organized regime ($p = 0.2$). (From B. Derrida, "Phase Transitions in Random Networks of Automata," in *Chance and Matter*, eds. J. Souletie, J. Vannimenus, and R. Stora, Les Houches 1986, North Holland 1987).

10–3–6 The "Magnetization"

The *magnetization* of a node, named in reference to the physics of magnetic media, is the temporal average of the state of the node. It is evaluated after the transient period. If the state of the node is fixed, the magnetization is 0 or 1. For short local periods, the magnetization takes on fractional values: $1/4, 1/3, 1/2, 2/3, \ldots$. Consequently, in the organized regime, the histogram of the observed magnetizations over the entire network is a series of discrete peaks (Fig. 10.13). On the other hand, in the chaotic regime, this histogram is continuous (Fig. 10.14). This behavioral difference can be compared to the frequency spectra of systems of nonlinear differential equations: periodic regimes are characterized by discrete spectra, whereas turbulent regimes give rise to continuous spectra.

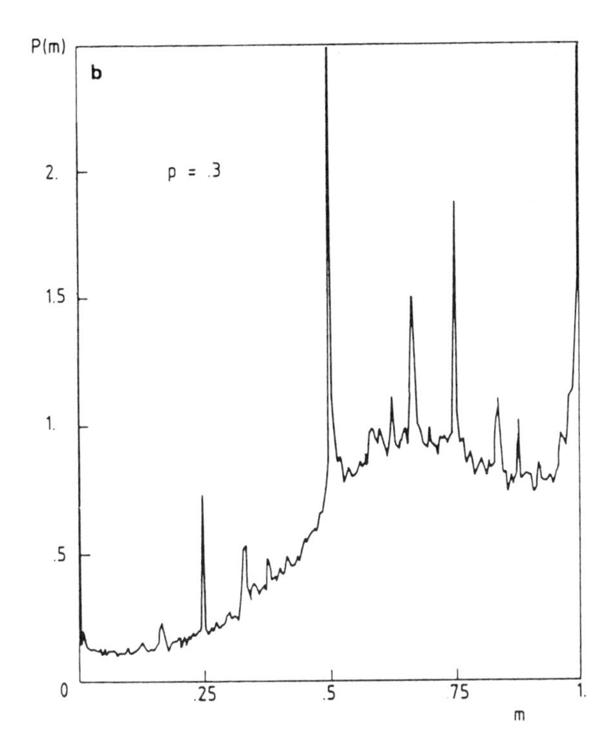

FIGURE 10.14 Histogram of the magnetizations in the chaotic regime ($p = 0.3$). "Phase Transitions in Random Networks of Automata," in *Chance and Matter, ibid.*).

10–4 The Distance Method

This method has recently been found to be one of the most fruitful techniques for determining the dynamics of a network. Recall that the Hamming distance between two configurations is the number of automata in different states. This distance is zero if the two configurations are identical, and equal to the number of automata if the configurations are complementary. We obtain the relative distance by dividing the Hamming distance by the number of automata.

The idea of the distance method is the following: we choose two initial conditions separated by a certain distance, and we follow the evolution in time of this distance. The quantity most often studied is the average of the asymptotic distance, measured in the limit as time goes to infinity. We compute this average over a large number of networks and of initial conditions, for a fixed initial distance. Depending on the initial distance, the two configurations can either evolve toward the same fixed point (in which case the distance goes to zero), or toward two different attrac-

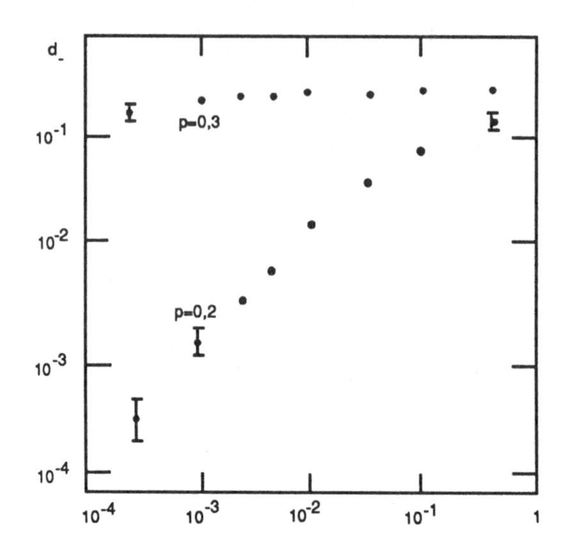

FIGURE 10.15 Relative distances at long times as a function of the initial relative distances, in the organized ($p = 0.2$) and chaotic ($p = 0.3$) regimes. (From B. Derrida and D. Stauffer, *Europhys. Lett.* **2**, p. 739, 1986.)

tors, or they could even stay a fixed distance apart (in the case of a single periodic attractor), regardless of whether the period is long or short. Again, we observe a difference in the behaviors of the two regimes. On figure 10.15, the x-axis is the average of the relative distances between the initial configurations, and the y-axis is the average of the relative distances in the limit as time goes to infinity. In the chaotic regime, we observe that if the initial distance is different from 0, the final distance is greater than 10%. The final distance seems almost independent of the initial distance. On the other hand, in the organized regime, the final distance is proportional to the initial distance.

We can in fact consider two neighboring configurations to be the set composed of a reference configuration and a configuration perturbed in a few places. The result obtained for the organized regime can easily be interpreted by taking the independent subnetworks into account: the final distance is proportional to the number of perturbed subnetworks, which is in turn proportional to the number of automata modified initially. In the case of the chaotic, unstructured regime, the perturbation propagates through the entire network. We therefore observe the sensitivity to initial conditions characteristic of chaotic regimes in continuous systems.

Diluted Networks

It is possible to predict the evolution of the distances in the case of diluted networks, .i.e., randomly connected networks for which the k/N ratio is close to 0.

First consider two configurations chosen at random in the case of an arbitrary network of connectivity k. We can evaluate the evolution of the overlap $x(t)$ in one iteration step. The overlap is the ratio of the number of automata in the same state to N. In other words:

$$x = 1 - d \tag{10.6}$$

where d is the relative distance. It is also the probability that two automata are in the same state. Initially, $Nx(t)$ automata are in the same state. Consequently, all the automata for which the k input automata are in this subset will stay in the same state at time $t+1$, after the first iteration. There are $Nx(t)^k$ of them. If the states and the automata are uncorrelated, all of the other automata have a fifty percent chance of staying in the same state. Consequently,

$$x(t+1) = \frac{1 + x^k(t)}{2}. \tag{10.7}$$

This expression is only valid for two random configurations. In principle, then, it cannot be iterated to determine x for long times, since the configurations attained in the course of the iteration are no longer random. In particular, the input states of a given node can be correlated by the existence of common ancestors. Equation (10.7) can nonetheless be iterated in the following cases:

- For diluted networks, for a limited number of steps so that the common ancestors of the inputs of a node are rare. This is true as long as the number of iterations is smaller than $\log N / \log k$.

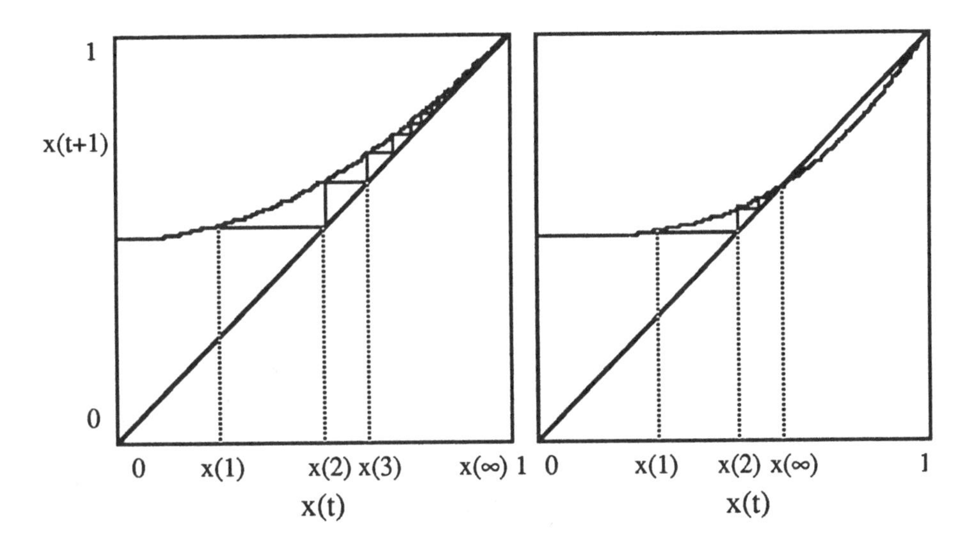

FIGURE 10.16 Iteration of the relative overlaps for connectivities of 2 (on the left; $x(t+1) = (1 + x^2(t))/2$) and 3 (on the right; $x(t+1) = (1 - x^3(t))/2$).

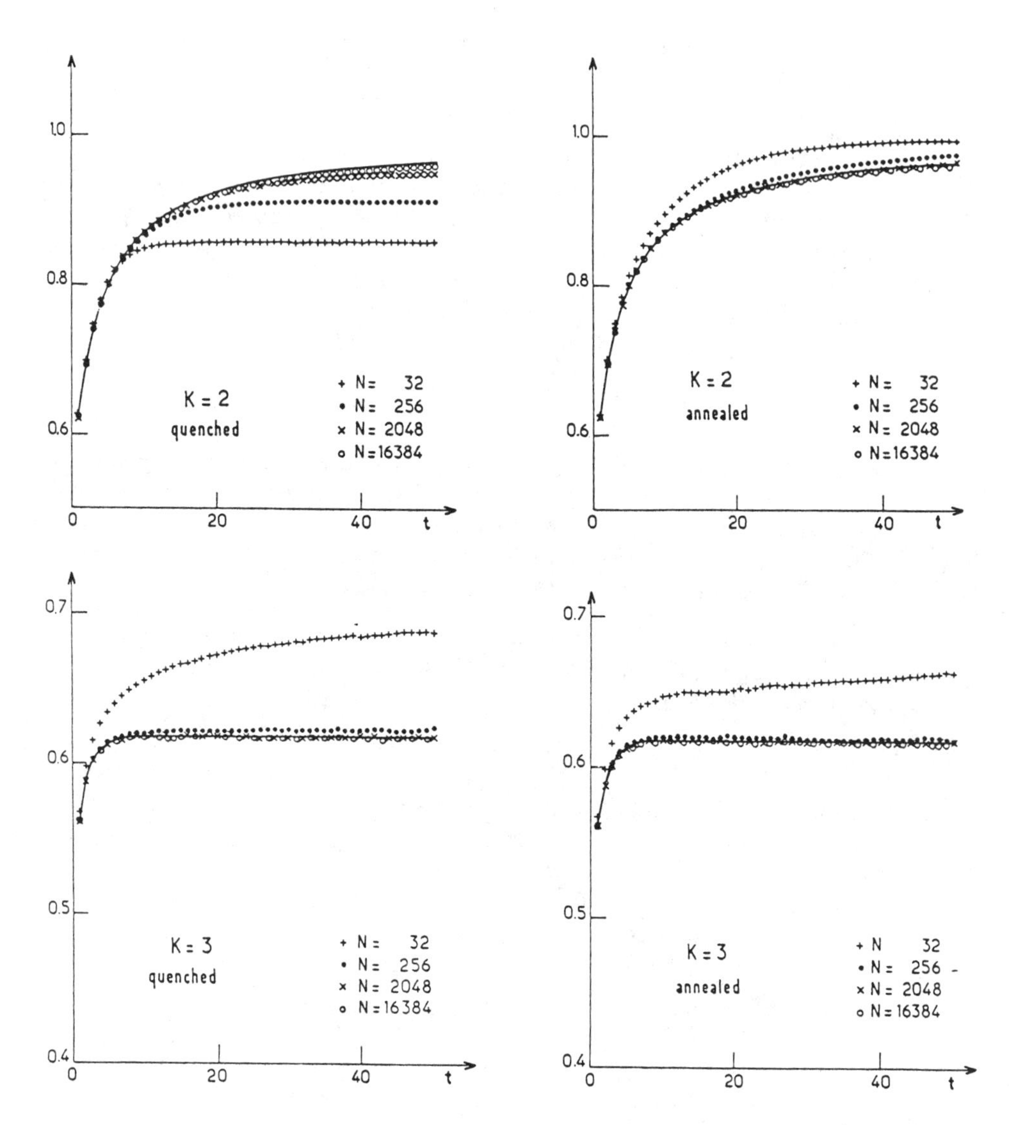

FIGURE 10.17 Evolution of the overlaps $x(t)$. Comparisons between "frozen" networks, on the left, and "annealed" networks, on the right. The figures on the top were obtained for connectivities $k = 2$, and the ones on the bottom for connectivities $k = 3$. The graphs in solid lines were obtained by iterating equation 10.7, and the points are the results of numerical simulations (+'s correspond to a network of 32 automata, ·'s to 256, x's to 2048, and o's to $16,384$). (From B. Derrida and G. Weisbuch, *J. Physique*, vol. 47, 1297 (1986).)

- For *annealed networks*. An annealed network is a network in which the functions of the automata are picked at random at each iteration, which destroys the correlations between the configurations and the automata. These networks were proposed by B. Derrida and Y. Pomeau.

If we iterate equation (10.7), we observe two different behaviors, depending on whether or not the graph of $x(t+1)$ as a function of $x(t)$ crosses the bisector of the plane for $x(t)$ in the $[0,1]$ interval. For small connectivities, up to $k=2$, there is no intersection in this interval and the overlap goes to 1 as t goes to infinity. This can be observed on figure 10.16. On the other hand, if $k>2$, the graph intersects the bisector before $x=1$, and so the overlap remains less than 1 as t goes to infinity.

In the first case shown in the figure, the distance between the two initial configurations goes to 0, which means that the number of configurations accessible to the system goes to 0 as t goes to infinity. This does indeed correspond to an organized behavior.

In the second case shown in the figure, the distance remains finite: the space which is accessible to the system is of the order of 2^{1-x}. In the limit of large t, the distance seems to be independent of the initial distance. This does indeed correspond to a chaotic behavior.

We have compared the theoretical predictions of equation 10.7 with numerical simulations on annealed networks and "frozen" networks, in other words networks for which the functions of the automata are fixed. We can see that the two types of networks behave in the same way at the start of the iterations. The phase transition predicted by equation 10.7 for annealed networks is the same as the one observed for frozen networks. The significant changes in behavior are mainly observed for large times for organized networks ($k=2$): the overlap goes to finite values less than 1 for frozen networks. However, the difference between this limit and 1 goes to 0 as the size of the network increases (Fig. 10.17).

10–5 Conclusions

This study clearly demonstrates the existence of two types of behaviors, organized and chaotic. Table 10.1 summarizes the differences in the generic properties of these two regimes.

The distance method has been used in a large number of cases to study the dynamical behaviors of different networks—networks with cellular or random connectivities, with threshold automata or boolean automata, or with probabilistic or deterministic automata. In the case of random diluted systems, it allows theoretical predictions to be made, such as predictions about the transition between regimes of boolean networks, or the capacity of diluted networks of threshold automata (see section 6.3). In the case of cellular connectivity, theoretical predictions are impossible, but the behaviors are qualitatively similar to those of diluted networks. In

TABLE 10.1 Generic properties of random networks

Property	Organized regime	Chaotic regime
Period	small	large
Scaling law (periods)	goes as a root of N	exponential in N
Oscillating nodes	isolated subnetworks	percolate
Magnetization	discrete spectrum	continuous spectrum
Evolution of distances		
Cellular networks	proportional to d_0	d_∞ finite, independent of d_0
Random Connectivity	d_∞ goes to 0	d_∞ remains finite.

addition, the technique is simple to implement in numerical simulations, and it allows the transition parameters to be determined relatively efficiently. The following section shows the application of this technique to spin glasses with nearest-neighbor interactions (Ising model, see section 8.5).

10–6 Application of the Distance Method to Spin Glasses

The network is defined by the interaction constants between neighboring spins: the T_{ij} are equal to $+1$ or -1 and are randomly chosen. The two configurations whose distances we are studying undergo parallel iterations. In addition, the dynamics is probabilistic in order to take the temperature into account, but the two configurations are subjected to the *same thermal noise*. Specifically, to choose the $+1$ state for spin i with a probability p given by equation (8.8), we compare p with a random number z between 0 and 1. If p is less than z, which occurs with a probability p, the spin takes on the value $+1$; otherwise, it takes on the value -1. The same thermal noise is applied to both configurations by using the same number z.

First of all, it is interesting to study the spatial organization of a spin glass, as we have done for networks of boolean automata. Consequently, we monitored the evolution of a two-dimensional network of 100×100 spins, at a temperature of 0.9. The left side of figure 10.18 shows the absolute value of the average magnetization after 1250 iterations, for configuration A. The white regions correspond to average magnetizations of zero: in these regions, the spins are perpetually oscillating. In the black regions, the spins have an amplitude of magnetization, S^2, of 1: they are therefore fixed.

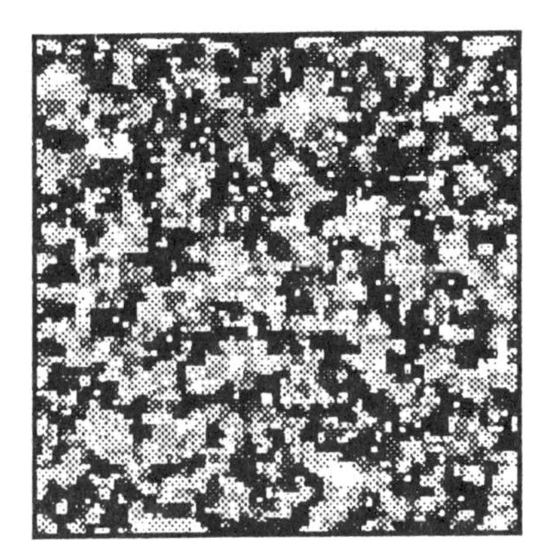

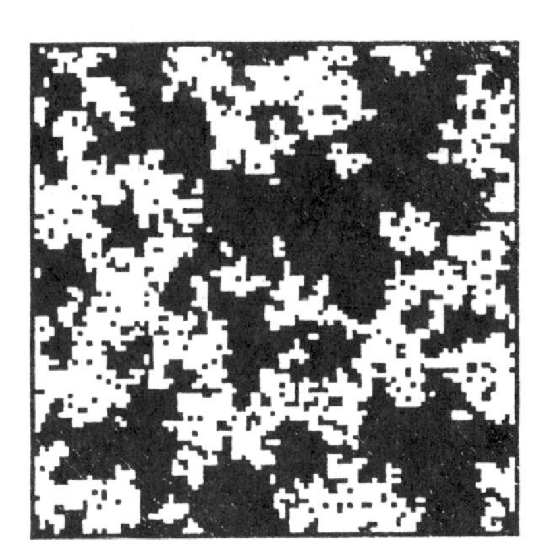

FIGURE 10.18 Domains in spin glasses. Two-dimensional sample of 100×100 spins, at a temperature of 0.9, after 1250 interations. The figure on the left is a map of the mobility of the spins: the light regions correspond to fluctuating spins, and the black regions to fixed spins. The figure on the right is a map of the distances between the spins for two configurations which at the start were randomly chosen: in the black regions, the spins are identical, and in the white regions, they are opposite. (From A. Neumann, B. Derrida, and G. Weisbuch, *Complex Systems* (1988)).

The figure on the right shows larger structures. It represents the distance between two configurations A and B which evolved in parallel, subjected to the same thermal noise. In the black regions, the spins are identical, and in the white regions, they are opposites. The attractive configurations are therefore divided into relatively independent domains which can potentially reverse their polarizations. Of course, if we directly observe each configuration, no large-scale structures are visible; because of the disorder, the states of two neighboring spins appear to be uncorrelated.

We can also take the average distance between two configurations over the entire sample. Figure 10.19 shows the evolution of the distance for large times ($t = 500$) as a function of temperature for cubic samples of $12 \times 12 \times 12$ spins. Three different behaviors can be seen:

- At high temperatures, $T > 4$, in the paramagnetic phase, there is a single attractor and the distance between the two configurations is zero.
- At low temperatures, $T < 1.8$, in the spin-glass phase, there are so many attractors that each configuration is trapped in an attractor near the initial configuration: it follows that the final distance between the configurations depends on the initial distance.

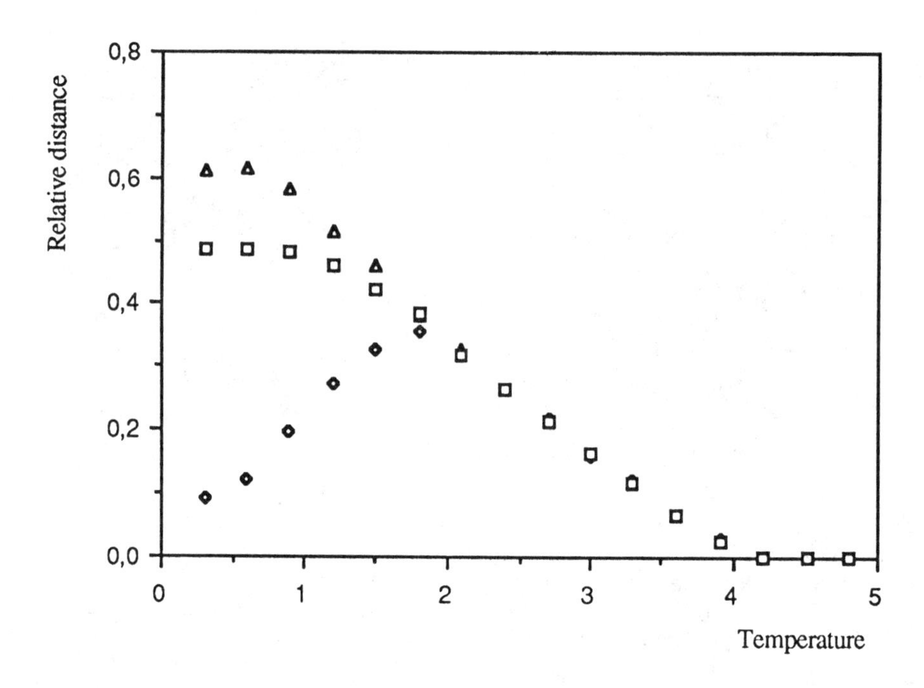

FIGURE 10.19 Relative distances as a function of temperature between two configurations of the same spin glass, after 500 iterations. The three-dimensional network contains $12 \times 12 \times 12$ spins. (The results were averaged over 200 samples.) The triangles correspond to two configurations which were initially opposite, the squares to two random configurations, and the diamonds to two configurations which initially differed in only one spin. For temperatures greater than 2, the graphs for the three initial distances are almost identical. (From B. Derrida and G. Weisbuch, *Europhysics Let.*, vol. 4, 657 (1987).)

- For intermediate temperatures, the final distance is always finite and independent of the initial distance. The attractors form large domains which hardly fluctuate as long as the temperature is not too high. Each configuration evolves toward one of these attractors, and the observed distance corresponds to the average distance between these attractors.

References

A good reference on the subject of the transition is "Dynamical Phase Transitions in Random Networks of Automata," by B. Derrida, in *Chance and Matter*, edited by J. Souletie, J. Vannimenus, and R. Stora (North Holland, 1987).

On the concept of percolations, see *Introduction to Percolation Theory*, by D. Stauffer (Taylor and Francis, 1985); and for chaos, *Order within Chaos*, by P. Bergé, Y. Pomeau, and C. Vidal (Wiley, 1986).

ELEVEN

Genotypes and Phenotypes

What are the principles of the functional organization of living systems derived from the genetic information carried by the genome?

This problem is often downplayed by classical molecular biology: from that standpoint, given that all the information is coded in the genome, the development of the organism is univocally "programmed." The same is true of the evolution of species, subject to a selection pressure: only the "fittest" survive. Along the same lines, the object of biology as a science is to describe the specific mechanisms of each interaction necessary to determine the behavior of the system being studied at every level, ranging from the conformation of macromolecules, to cellular metabolism, or ontogenetic or phylogenetic development.

Conversely, in this chapter we will attempt to describe the emergence of collective properties of organization independently of the details of the interactions between the components. In other words, we are in search of generic behaviors, in the sense defined in the previous chapter.

The genome, or *genotype*, is the genetic message contained in the DNA of the chromosomes. It is the genome which determines, partly as a function of external conditions, the *phenotype* of the organism, that is to say the set of its physical and chemical properties, as well as its biological functions, such as the arrangement and function of the organs, height, weight, sex, etc. But this determinism, accepted by all scientists, is far from being understood at all levels. What is clear today is the genetic code, i.e., the correspondence between the bases of the nucleic acids (their elementary components) and the amino acids which form proteins. However, as soon as we go beyond this first level, mysteries arise: How does the sequence of amino acids determine the spatial configuration of proteins, which is responsible for their chemical reactivity? How do the proteins interact to determine the cellular structure? And so on. The series of questions continues up to the level of the development of organs, of the organism, and of the interactions between populations. In this chapter we will tackle the genotype/phenotype question by using an automata network approach and concerning ourselves with very general properties of living beings. We will discuss three examples: cellular differentiation, the origin of life, and the evolution of species.

11–1 Cell Differentiation

The apparent paradox of cell differentiation is the following:

> "Since all cells contain the same genetic information, how can there exist cells of different types within a single multicellular organism?"

Indeed, our body contains cells with very different morphologies and biological functions: neurons, liver cells, red blood cells... a total of more than 200 different cell types. Yet the chromosomes, which carry the genetic information, are the same in different cells. Part of the answer is that not all of the proteins coded for by the genome are expressed (synthesized with a non-zero concentration) in a cell of a given type. Hemoglobin is found only in red blood cells, neurotransmitters and their receptors only appear in neurons, etc.

Several mechanisms can interfere with the different stages of gene expression to facilitate or block it. We speak of activation and repression. The best known mechanisms involve the first steps of transcription. In order to transcribe the DNA, a specific protein, DNA polymerase, must be able to bind to a region of the chain, called the promotor region, which precedes the coded part of the macromolecule. Now, this promotor can be partially covered by a control protein, called the repressor; reading the rest of the chain is then impossible. It follows that, depending on the quantity of repressor present, the gene is either expressed or not expressed. The protein which acts as a repressor is also coded for by another gene, which is itself under the control of one or several proteins. It is tempting to model the network of these interdependent interactions by an automata network.

- A gene is then represented by an automaton whose binary state indicates whether or not it is expressed. If the gene is in state 1, it is expressed and the protein is present in large concentrations in the cell. It is therefore liable to control the expression of other genes.
- The action of control proteins on this gene is represented by a boolean function whose inputs are the genes which code for the proteins controlling its expression.
- The genome itself is represented by a network of boolean automata which represents the interactions between the genes.

In such a network, the only configurations which remain after several iteration cycles are the attractors of the dynamics, which are fixed points or limit cycles, at least when the dynamics is not chaotic. These configurations can be interpreted in terms of cell types: a configuration corresponds to the presence of certain proteins, and consequently to the biological function of a cell and its morphology. Consequently, *if* we knew the set of control mechanisms of each of the genes of an organism, we could predict the cell types. In fact, this is never the case, even for the simplest organisms. We only know the control mechanisms in the case of a few genes. The sct of mechanisms leading to the consumption of lactose in *Escherichia coli*, called the lactose operon, has been described by Monod et al. Without knowing the complete diagram of the interactions, S. Kauffman (1969) set out to uncover the generic properties common to all genomes by representing them by random boolean networks. Since there is a finite number of possible boolean laws for an automaton with a given input connectivity k, it is possible to construct a random network with a given connectivity.

As was stated in the preceding chapter, by using numerical simulations, Kauffman determined the scaling laws relating the average period of the limit cycles and the number of different limit cycles to N, the number of automata in the network. For a connectivity of 2, these two quantities seem to depend on the square root of N (in fact the fluctuations are very large). In fact, these same scaling laws have been observed for the time between cell divisions and for the number of cell types as a function of the number of genes per cell (Fig. 11.1). The number of different cell types matches the number of attractors in Kauffman's model quite well. As for the time between cell divisions, it makes sense to compare it to the temporal variable related to the attractors of the dynamics, the period, in this case.

It is clear that Kauffman's approximations are extremely crude compared to the biological reality—binary variables representing protein concentrations, boolean (and thus discrete) functions, simultaneity of the transitions of automata, random structures.... The robustness of the results obtained with respect to the possible modifications of the model (these are random networks) justifies this approach. As for the existence of a large number of attractors, it is certainly not related to the particular specifications of the chosen networks; it is a generic property of complex systems, which appears as soon as frustrations exist in the network of the interactions between the elements.

11–2 The Origin of Life

The questions regarding the origin of life on Earth are different depending on the level of organization we are interested in. Historically, the problem was first stated in terms of chemistry: how can we explain the appearance of organic molecules in an oxidizing medium in which these components would have a greater tendency to burn, releasing energy? The focus then shifted to functional groups more directly related to life, such as amino acids and nucleotides. A series of experiments were begun in the fifties, which were somewhat reminiscent of the motivations and

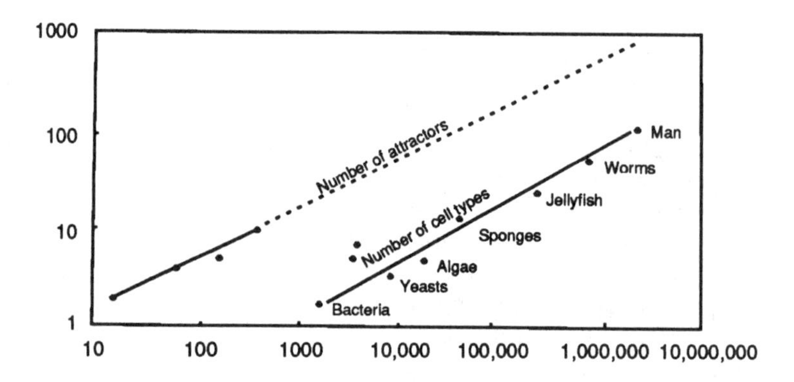

FIGURE 11.1 Graph of the number of cell types as a function of the number of genes per cell. Here the two scaling laws are shown: the one obtained for Kauffman's model, which corresponds to the number of different limit cycles as a function of the number of automata, and the one for biological organisms. Both scales are logarithmic. The two graphs are approximately parallel with slopes of about 0.5. (From S. Kauffman, "Binary Elements Nets" in *Towards a Theoretical Biology 3. Drafts*, edited by C. Waddington (Edinburgh University Press, 1970).)

the methods of Berthelot's egg. Different elementary inorganic components, such as water, air, and carbon gases, were mixed together in a test tube; and the presence of "interesting" components such as amino acids and nucleotides were observed, after subjecting the medium to an electrical discharge (Orgel).

In fact the most delicate conceptual problem is at the biochemical level and concerns the replication of the nucleic acids and protein synthesis. Indeed, both operations require the presence of proteins which act as specific catalysts. Before the "appearance" of proteins, protein synthesis as well as the synthesis of nucleic acids would therefore have been impossible, even though they are necessary for this synthesis to occur. In short, this is the biochemical version of the chicken and the egg problem. The answer in terms of physics is obviously to invoke a mechanism of instability.

First of all, in Prigogine's terminology, one must imagine autocatalytic mechanisms during which the reaction product is used in some of the reactions which occur during its synthesis. The first model of the appearance of autocatalytic "hypercycles" between proteins and nucleic acids was cast in this framework by M. Eigen (1971). It is based on a system of nonlinear first-order differential equations for the chemical concentrations (C_i) of the $(i, j, k \ldots)$ species in the solution. One of these kinetic equations, for example, is

$$\frac{dC_i}{dt} = (a_i + b_{ij}C_j)C_i + c_{ik}C_k \tag{11.1}$$

where the $b_{ij}C_jC_i$ term represents the synthesis of the chemical species i catalyzed by species j. Since the b_{ij} are positive, these systems of nonlinear differential equations have unstable modes, and the one for which the growth is the fastest wins over the others. A set of components and a reproductive metabolism between these selected species then "emerges."

The dynamics proposed by Eigen is rather simplistic in that it is entirely deterministic (there is no error in the duplication of the DNA), and that it assumes that all of the components which are liable to interact are present in the solution in finite concentrations. It can only lead to a single hypercycle, the most unstable one, which is somewhat at odds with the diversity of organisms observed in nature. On the other hand, it guarantees the stability of this hypercycle.

In an attempt to reproduce both the stability and the diversity observed in the living world, P. Anderson proposed a model based on an analogy with spin glasses. The model is at the level of the individual molecules, and not at the level of their concentrations, which makes the numerical simulations longer and more difficult.

Consider at the outset a "soup" composed of monomers, of a few oligomers (small chains of a few monomers), and of energetic molecules capable of allowing the biochemical reactions. We simplify the biochemical problem by assuming that initially, the proteins were not necessary to maintain the autocatalytic cycles between the polynucleotides. We can safely make the further simplification that there are only two different nucleotides, for example A and T, represented by the spins $+1$ and -1. The sequence of bases of a polynucleotide is therefore represented by a linear chain of spins $+1$ and -1.

The polynucleotides exist in solutions in the form of simple linear chains, or in the form of double helices with two chains, associated by the complementary base pairs $+-$ or $-+$ which are connected by weak hydrogen bonds. The transition between the two forms is reversible. Low temperatures and high concentrations favor the double helix form, whereas high temperatures and low concentrations favor the separation of the helix into simple chains.

This leads us to consider two types of chemical reactions.

- The first sequence of reactions involves the *synthesis* of the polynucleotides. This synthesis involves a chain of three reactions, where the two reactions on the ends are the inverses of each other, and presuppose cyclical variations in the environment. It leads to the concatenation on a "template" of pieces of chains complementary to the template, in the following three steps:

First step: formation of a double helix as a consequence of a change in physical conditions (such as a lowering of the temperature). This double helix is composed of a chain which serves as a template, and of two other chains, partly complementary to the template, which affix themselves to it by hydrogen bonds between complementary bases (see Figure 11.2).

Second step: formation of a covalent bond between the two ends of the pieces which have bonded to the template.

FIGURE 11.2 Two chains (top) that are partly complementary to the template (bottom) affix themselves to it by hydrogen bonds between complementary bases.

Third step: the two chains again separate, subject to heating, for example, and are then available to intervene in a new cycle, either as templates or as reagents. This type of reaction includes reproduction with a potential for mutations, since it is not necessary for the chains to be perfectly complementary in order to form a double helix. Of course, the binding of a monomer at the end of the chain is but a special case of this mechanism. The basic hypothesis on the alternation of hot and cold periods can be easily explained by cycles of light and shadow or by the cycles of day and night. This series of reactions has already been proposed and studied by biochemists (Blum 1962, Kuhn 1978, etc.), and the contribution by theoretical physicists relates to the dissociation reactions.

- The *dissociation* reactions of a chain.

In a given chemical environment, there always exist hydrolysis reactions of a chain which lead to its dissociation. In a kinetic chemical model such as M. Eigen's, for example, this is expressed by a a_iC_i term with a_i negative. In a numerical simulation, such as the one by P. Anderson, D. Stein, and D. Rokhsar, this is expressed by introducing a random event with a dissociation probability. This probability will depend on many factors, such as the strength of the covalent bonds. If the bonds between neighboring bases were independent from one another, the dissociation probability of the chain would involve a product of the dissociation probabilities of each bond; the problem could then be reduced to a statistical mechanics problem between independent particles. However, the accessibility of the bonds depends on the secondary structure (the hydrogen bonds between the bases of the same chain), as well as the three-dimensional structure of the chain. This is therefore a *collective property* of the set of bases of the chain. From this stems the idea of calculating this probability from an expression similar to that of the energy in a spin glass. The dissociation function of a chain is given by:

$$D = \sum_{i,j} T_{ij} S_i S_j \tag{11.2}$$

where the subscripts i and j refer to the monomers.

The dissociation probability used in numerical simulations is a Fermi-Dirac function of the dissociation function:

$$P = \frac{1}{1 + \exp[-D - m(N)]} \tag{11.3}$$

where $m(N)$, the chemical potential, maintains the length of the polymers close to a certain average.

Since we don't know the T_{ij} involved in the physical chemistry of the polynucleotides, we choose them according to a random distribution, as in the Sherrington-Kirkpatrick model, for example.

Of course, the chemical species which have the smallest dissociation function are favored. The reproductive mechanism by complementary base pairs described above, along with the natural selection guaranteed by the dissociation function, result in the stability of a certain number of chemical species. The diversity of these species is guaranteed by the existence of a large number of secondary minima of the dissociation functions.

The numerical simulations start from a primeval soup where only monomers and all of the possible trimers are present in equal concentrations, so as to not favor the growth of any particular sequence *a priori*. They alternate between phases of concatenation onto templates, involving searches for pieces of chains capable of associating themselves onto the chain by complementarity; and phases of dissociation of the chains by random trials involving comparisons with the dissociation probability of the chain.

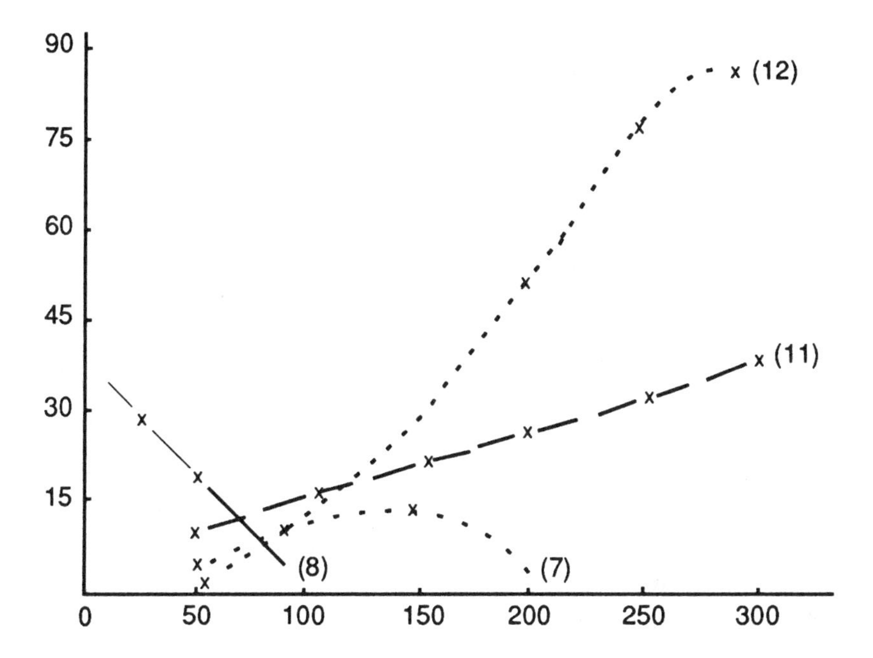

FIGURE 11.3 Evolution of the populations of different interacting nucleotides in a primeval soup. We can see that two of the four monitored species are selected, whereas the two others disappear. (From D. Rokhsar, D. Stein, and P. Anderson, *J. Mol. Evol.*, vol. 23, 119 (1986).)

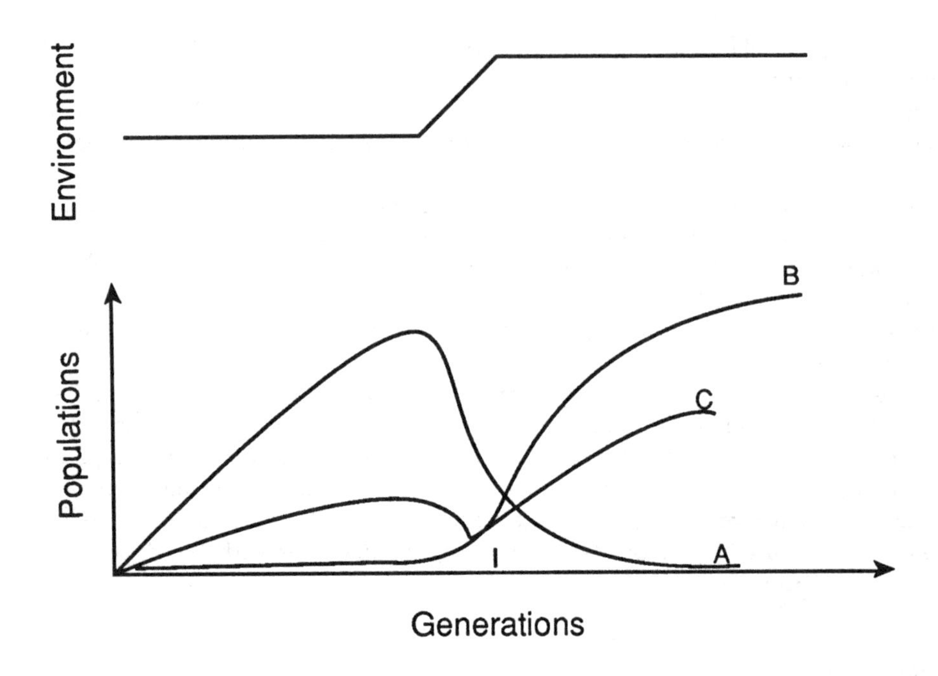

FIGURE 11.4 Evolution of three populations of nucleotides subject to a change in the environment. Species A is no longer fit, and disappears. Species B benefits from the change. Species C adapts after mutating. (From D. Rokhsar, D. Stein, and P. Anderson, *J. Mol. Evol., ibid.*)

Figure 11.3, for example, shows the disappearance of certain species and the growing predominance of others over time.

Figure 11.4 shows the effect of a change in the environment, modeled as a modification of the T_{ij}. Species A disappears, species B invades the space, and species C, after some initial difficulties, adapts after undergoing a mutation.

Numerous simulations which have come out of the Princeton group have confirmed the generic character of the results obtained. For a large range of interaction parameters and of chemical potentials, we find qualitatively similar results: the selection of a small number of species corresponding to the valleys of the D function, and ultimately the stability of the selected species. This leads to the idea that the observed behaviors are independent of the details of the model. In particular, if instead of a spin-glass-type function, we could use the experimental biochemical data, we would observe the same qualitative behaviors.

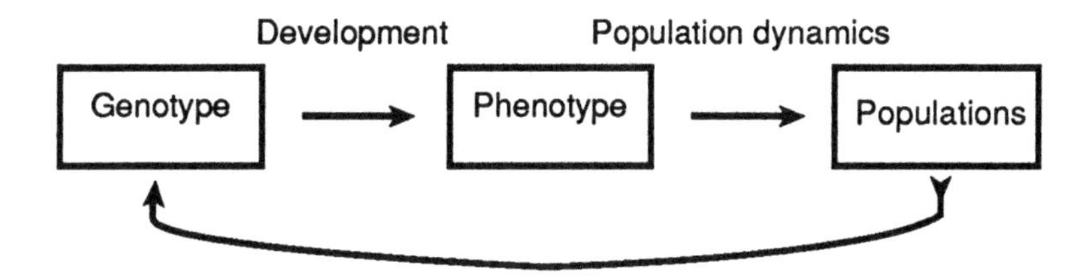

FIGURE 11.5 The interaction loop between the three levels involved in the evolution of species.

11–3 The Evolution of Species

The evolution of species is a three-level dynamical process: genotype, phenotype, and populations. As for the origin of life, an additional complication is introduced in the dynamics by a feedback loop between the highest level, that of the populations, and the lowest level, that of the genome (Fig. 11.5).

Is it possible to characterize the generic properties of the dynamics despite our ignorance of the mechanisms of development (going from genotype to phenotype) and of "natural" selection (going from phenotypes to populations)? The timeliness of this question is in part related to an ongoing debate on "punctuated equilibria" started by the two paleontologists S. Gould and N. Eldredge.

There exist several semi-quantitative results on the dynamics of evolution. Unfortunately, these results seem contradictory, depending on whether one studies the genomes or the phenotypic properties. The studies on genomes involved comparisons of the homologous proteins (i.e., proteins having the same biological function) of different living species (Zuckerkandl). The conclusions are that the rate of mutations, measured by the differences between sequences of amino acids, has stayed constant throughout the history of the planet. On the other hand, if we measure certain phenotypic variables such as the size of the organisms, we sometimes observe *punctuated equilibria*: long stases during which evolution seems to have stopped, are separated by short periods of very rapid evolution, which appear as discontinuities in the fossil record.

Standard mathematical models of evolution, such as population genetics models, are based on an extreme "Mendelian" hypothesis—the one gene/one character correspondence. In other words, each phenotypic character is determined by a single gene. In this framework, a constant mutation rate at the level of the genes cannot lead to abrupt jumps such as punctuated equilibria at the phenotypic level. To solve this paradox, paleontologists can invoke two different arguments:

- The occurrence of isolated events, such as "catastrophic" climatic or geological changes, or meteorites, which could have suddenly changed the living conditions on Earth.

- The possibility that the mutations responsible for the changes in the species are of a different type than those which are observed within a same species, and which are correctly described by standard population genetics.

In fact, we will see that if we take into account the fact that the phenotype is a global property of the genome, and not a collection of independent characters, each determined by a single gene, no additional hypothesis such as the ones proposed by the paleontologists is necessary to explain punctuated equilibria.

11–3–1 Population Dynamics

THE ROLE OF ROBUSTNESS Suppose that all of the possible organisms have a genome containing N genes, and that each gene can have four different forms, or alleles. Therefore there exist 4^N possible organisms. Let P_i be the population of organism i. In order to characterize evolution, we must describe the variations of the populations in time in the 4^N-dimensional space of all the genomes. At each point in this space, a population evolves under the influence of births, deaths, and competition between the organisms; subject to mutations, some organisms can occasionally jump from one point to another. We can imagine that the populations are governed by a system of differential equations containing terms for reproduction, death, etc. Once all of these terms are known, one need only solve these equations, or at least characterize the dynamics by making reasonable simplifications.

The essential difficulty of this program is to take into account the connection between the genome of organisms i and the terms of the equations. In order to do this, it is necessary to understand the mechanisms of development, and we are still very far from this. We will circumvent this difficulty by taking into account the fact that the phenotype involves the collective properties of the genotype. Suppose that the genome is modeled by a Kauffman network with small connectivity. The phenotypic characters then depend on the attractors of the network dynamics. Now, the global dynamical properties of this network are characterized by a high degree of *robustness*. The robustness stems from the fact that changing a small fraction of the automata in a network has very little effect on its attractors. Up to this point, we have only discussed the stability of the attractors with respect to small perturbations of initial configurations. But clearly, the stability with respect to initial conditions leads to robustness with respect to small perturbations of the network. The reasons are the same: forcing structures and structuring into independent subnetworks. Numerical simulations make it possible to verify the robustness of random boolean networks. The same robustness can also be observed for Hopfield or similar networks: in an attractive configuration, the potential failure of a single automaton does not modify the states of the other automata when the signal term is large. The robustness also indicates that two genomes which only differ in a few automata, and which are therefore neighbors in genome space, have neighboring dynamical properties.

THE INITIAL EQUATIONS We start from the system of differential equations 11.4, which describe the evolutions of the populations P_i at each point i in the space of genomes:

$$\dot{P}_i = a_i P_i - b \sum_j P_j P_i + \sum_v m(t) P_v \,. \tag{11.4}$$

The subscripts i, j, and v refer to the organisms. i is specific to the organism being considered, j can be any organism, and v represents organisms which differ from i in only a single gene.

The derivative with respect to time of the population of the organisms is therefore the sum of three terms.

- The first term is a growth term which takes into account the balance between reproduction, death of individuals, and mutations toward neighboring genomes. The proportionality coefficient a_i is called the *fitness* of organism i.
- The second term represents in differential form the sharing of resources, such as food, between all of the organisms; this is the reason for the sum over j.
- The third term represents the growth of the population due to the mutations from organisms of neighboring genomes. The amplitude of this term is proportional to $m(t)$, which fluctuates over time around an average rate m, which is small compared to the fitness coefficients.

This is of course an extremely simplistic model which does not take sexual reproduction into account, which only allows point mutations, and for which the competition between organisms is limited to sharing a single resource.

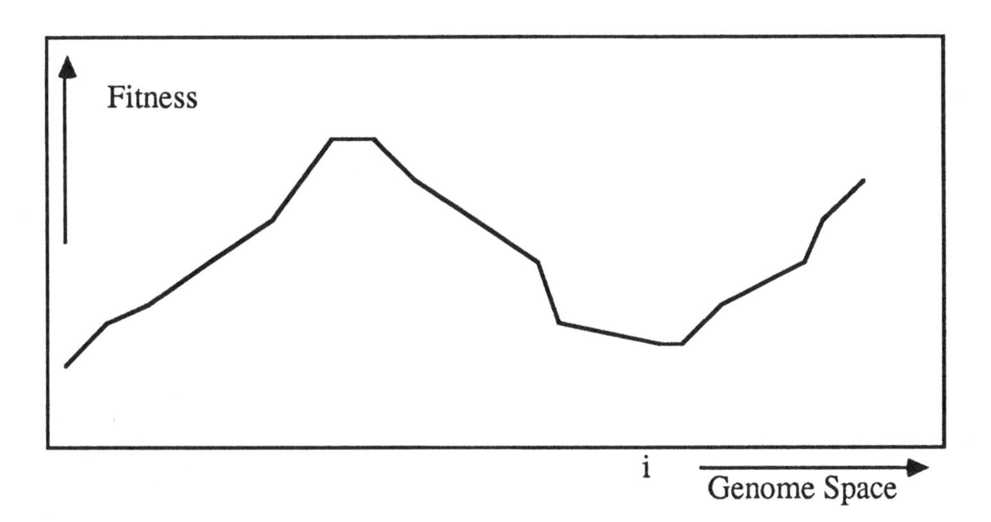

FIGURE 11.6 A robust fitness landscape.

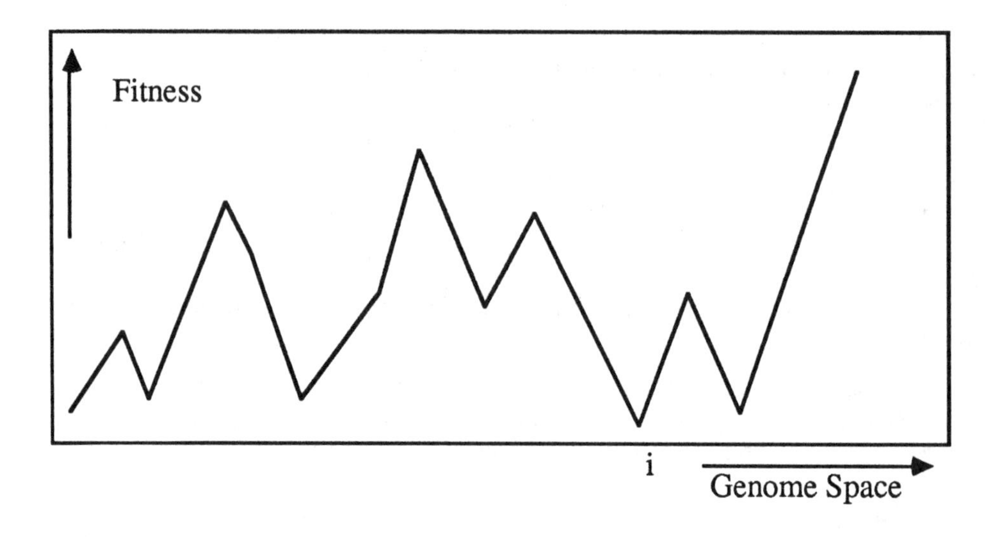

FIGURE 11.7 A random fitness landscape.

In fact, the basis of the model is that the fitnesses a_i can be chosen to be proportional to one of the dynamical properties of the network i, such as for example the longest period of the attractors, the average period, the number of attractors, etc. In other words, the fitness landscape in the space of genomes looks like figure 11.6 (robust landscape) rather than like figure 11.7 (random landscape).

Our conjecture, which is confirmed by simulations, is that any choice of a global property of the network to set the fitness leads, for the evolution of the species described by system 11.4, to a dynamical behavior of punctuated equilibria for small values of m.

SELECTION AND EVOLUTION If we were to start from an initial situation in which all of the species were present, the dynamics would be very simple. For the moment, let's assume that the mutations are negligible.

$$\frac{\dot{P}_i}{P_i} = a_i - b \sum_j P_j = a_i - bP_t \tag{11.5}$$

where P_t is the total population. We can see that the populations of fit organisms, that is to say with fitnesses greater than bP_t, increase. But the growth of the total population progressively renders unfit all of the organisms whose fitnesses are not maxima. Consequently, only those organisms with maximum fitnesses survive, and their populations eventually saturate. The other organisms disappear, except those whose genomes are neighbors of the fit organisms, because of the mutations.

In fact, the evolution of species corresponds to the case where, at the beginning of time, only a few populations exist on Earth, and none of them is most fit. Because of mutations, other organisms appear, whose successes, i.e., populations, depend on their fitnesses compared to the others, the available resources, and the history which might have favored the appearance of neighboring genomes. In dynamical terms, we are essentially interested in transient phenomena, or metastable states, rather than the search for an equilibrium.

11–3–2 Solution of the System of Differential Equations by the Perturbation Method

THE GROWTH PHASE Suppose that at time $t = 0$ a single organism is present. Its population increases, as well as the populations of neighboring genomes populated by mutations. Those neighbors of greater fitnesses grow more rapidly and their population is soon greater than that of the original ancestor. For small values of m, a narrow cloud formed by the most fit genomes moves in the space of genomes toward a local fitness maximum. This dynamics is very fast. If m is smaller than b, the mutation of a predecessor into a fitter mutant takes place after the population of the predecessor has saturated, and the time between mutations is given by:

$$t = \frac{\ln(m/b)}{a_i} + \frac{b}{ma_i}. \tag{11.6}$$

For the intermediate values of m for which useful mutations happen before saturation, the interval between mutations is given by:

$$t = \frac{\ln(m/a_i)}{a_i}. \tag{11.7}$$

In both cases we observe an accelerated evolution: the time interval between mutations decreases as the fitness increases.

THE STASIS When the local maximum has been reached, a quasi-equilibrium is established. The populations of the fittest organisms saturate, and those of the neighboring organisms are sustained by mutations from the fitness peak. The perturbation method allows us to calculate each of the populations as we move progressively farther away from the peak. This method is reminiscent of a Taylor series expansion, where m is the small term. At equilibrium, we assume that all the left-hand sides of the equations are zero, and in the right-hand sides we keep only the terms of smallest degree in m. We then obtain a series of relations:

$$\frac{dPm}{dt} = a_m P_m - bP_m P_m = 0 \tag{11.8}$$

$$P_m = \frac{a_m}{b} \tag{11.9}$$

where the subscript m refers to the local maximum.

For the nearest neighbors, denoted by the subscript v:

$$P_v = \frac{mP_m}{a_m - a_v} \tag{11.10}$$

which can be generalized for the distance d, the number of point mutations between the maximum and the genome i:

$$P_i = P_m m^d \sum_c \prod_j \frac{1}{a_m - a_j}. \tag{11.11}$$

This sum is performed over all of the paths from m to i through the different j.

Since the fitness landscape is soft, which is a consequence of the robustness, the width of the population distribution is smaller than the distance between the local maxima. It follows that the valleys are empty, since the probability for mutants to appear there is small, and since they disappear rapidly because of their low fitnesses. The cloud of populations centered on the peak, isolated from the other organisms, constitutes a *species*, in other words a set of organisms with slightly different genomes. This notion of species appears as a consequence of robust fitness landscapes, independently of a reproduction discontinuity between two sexual partners which are too different.

The absence of populations at the bottoms of valleys is the reason for the existence of stases.

THE APPEARANCE OF NEW SPECIES For a favorable mutation, i.e., a mutation which is more fit than the organisms at the peak, to appear, it takes a time t given by:

$$t = \frac{1}{mP_m} \prod_i \frac{a_m - a_i}{m} \tag{11.12}$$

where the product is taken along the pass, or the line of steepest slope connecting the peak to the favorable mutant. After the appearance of the mutant, the population cloud leaves the old peak and drifts very quickly toward the new peak, climbing the slope toward the favorable mutant.

NUMERICAL SIMULATIONS We can simulate the evolution of populations of organisms whose genomes are represented by a boolean network with six automata-genes. The graph of the interactions is shown in figure 11.8.

We use only the four boolean functions NAND, OR, AND, and NOR, which we code 0, 1, 2, and 3, respectively. The total number of possible genomes is $4^6 = 4096$. We measure a dynamical property such as the longest period for each of these 4096

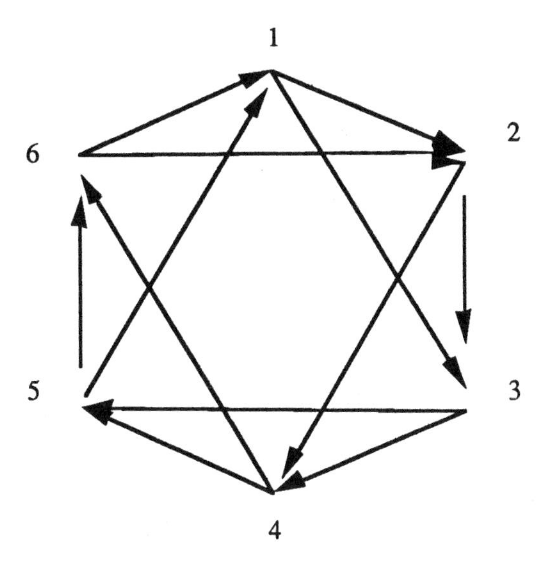

FIGURE 11.8 Connectivity structure of the boolean networks being studied.

networks. We only allow mutations which change one function at a time, restricting ourselves to the changes represented by the arrows as follows:

$$\text{AND} \leftrightarrow \text{NOR} \leftrightarrow \text{NAND} \leftrightarrow \text{OR} \leftrightarrow \text{AND}$$

In other words, the allowed mutations only change two bits of the output function. We can test the robustness of the period when we cover the space of genomes by a series of such mutations. Table 11.1 shows a cross section of this six-dimensional space. Each square represents a genome i, and the corresponding period T_i is written inside. The coordinates of the genomes are of the form $0023xy$, where x is the abscissa of the column, representing the function of automaton number 1, and y is the ordinate of the row, representing the function of automaton number 2. 0023 means that the automata 6, 5, 4, and 3 have the functions NAND, NAND, AND, and NOR, respectively.

The simulations whose results are shown in figure 11.9 were carried out for the following parameters:

$$m = 2 \times 10^{-5}, \qquad b = 10^{-7}, \qquad a_i = 0.05 \times T_i$$

At time $t = 0$, a single organism with a genome $i = 002332$ and a maximum period of 10 is present, on a local peak with a fitness a_i of 0.5. We can follow the evolution of the population of the initial genome as well as those of the genomes "sown" by mutations (of the order of ten). We observe the evolution toward peak 002300 in four steps characterized by different dynamical regimes:

TABLE 1 The longest period T_i in the $0023yx$ hyperplane of the genome space.[1]

	x				
y	2	3	0	1	2
2	7	1	4	1	7
3	10	1	3	4	10
0	4	4	11	9	8
1	1	1	1	1	1

[1] This table already illustrates the robustness of the period, even though the network is relatively small. The two peaks with coordinates 002332 and 002300, and of periods 10 and 11, are separated by three mutations (functions 0 and 3 are separated by a single mutation). The path of most favorable mutations goes through 002302, the pass, and 002301.

- The first step is the rapid exponential growth (about one hundred time steps) of the populations centered on the first peak. A first species is therefore composed of the population of the peak of period 10 and its two neighbors of periods 4 and 8.
- The saturated population stays centered on the peak for 85,998 time steps. This is the stasis. The mutant of period 9 appears and disappears.
- At time $t = 85{,}998$, the favorable mutant 002300 of period 11 appears for the first time. Its population then grows exponentially, becoming greater than that of the preceding species, which dies out because of the competition, becoming a fossil.
- Once the new equilibrium has been reached and the new species has been established, we once again enter a quasi-static regime.

We obtain similar results for a wide range of values for m, and for a different choice of the dynamical quantity which is used to determine the fitnesses. The peaks can then appear for different networks, but the semi-qualitative properties of the population dynamics, the alternation between rapid changes and stases, remain the same. The same would be true if we based the calculations of the adaptabilities on a spin-glass model such as Anderson's. The model is therefore very general; it is based on robust hypotheses and shows the same punctuated equilibria observed by paleontologists. This is a minimal model in the sense that to explain the facts, it does not require any new hypotheses beyond the classical neo-Darwinian theory.

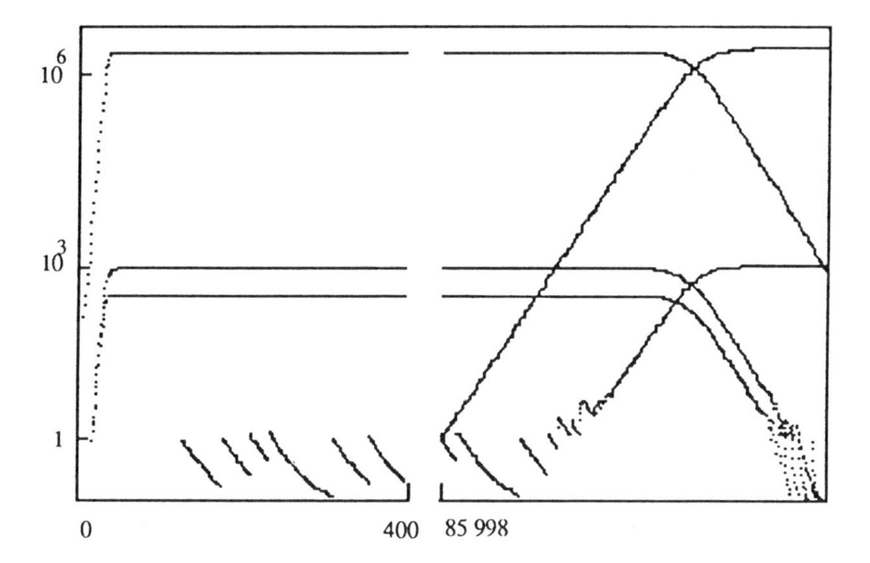

FIGURE 11.9 Numerical simulation of the evolution of populations of organisms with genomes containing six genes. The time is represented on the abscissa, and the ordinate is the $\log_{10}$ of the populations. The interval of time between 400 and 85,988 is not shown; during this time, the populations remain stable at the levels reached at t =399. For $t < 400$, the different graphs correspond to the genomes of periods 10, 8, 4, and 9, in the order of decreasing populations. At time t =85,998, a fitter organism with a period of 11 appears, and its population quickly surpasses all of the others.

In chapter 12, we will discuss the general impact of models based on automata networks. It can seem very ambitious to tackle problems as profound as the origin of life or the evolution of species with such simplistic models. In fact, none of the authors of the models discussed in this chapter claim to have settled the controversies surrounding these questions. The aim of these models is to show that as long as the interactions between the genes are taken into account, even for the most primitive models, we can obtain dynamical behaviors similar to those of living systems, which seem so astonishing to the biologists who observe them.

References

The reader who is a novice in the subject of molecular biology can refer to one of the classic texts such as *Molecular Biology of the Cell* by B. Alberts, D. Bray, et al. (Garland, 1989) or *Molecular Biology of the Gene*, by J. Watson, N. Hopkins, et al. (Benjamin and Cummins, 1989).

The book by J. Ninio, *Molecular Approaches to Evolution* (Princeton University Press, 1983), is a good description of the questions raised by the origin of life and evolution. S. Kauffman's book, *The Origins of Order: Self-organization and Selection in Evolution* (Oxford University Press, 1990), contributes some original ideas toward answering them.

Kauffman's model is described in S. Kauffman, *J. Theor. Biol.*, vol. 22, 437–467 (1969).

The hypercycle approach was proposed by M. Eigen, *Naturwissenschaften*, vol. 64, 541 (1971).

Anderson's model is described in P. Anderson, *Proc. Nat. Acad. Sc.*, vol. 80, 3386 (1983), and in D. Rokhsar, D. Stein, and P. Anderson, *J. Mol. Evol.*, vol. 23, 119 (1986).

A good reference on standard population genetics is W. J. Ewens, *Mathematical Population Genetics* (Springer Verlag, 1979).

The punctuated equilibria paradox appears in S. Gould and N. Eldredge, *Paleobiology*, vol. 3, 177 (1977), and the idea of a fitness landscape in S. Wright, *Evolution*, vol. 36, 427–443 (1982).

The conference proceedings *Artificial Life*, edited by Christopher Langton (Addison-Wesley, 1988, in this series), extends the discussion of the subjects in this chapter.

TWELVE
Conclusions

After having studied this book, the reader has certainly noticed the wealth of methods proposed to solve a wide range of problems. In this last chapter, we will attempt a provisional summary of this subject of ongoing research, pointing out a few of the bibliographic paths toward the numerous applications of automata networks. But first, it is perhaps appropriate now to comment on the scope of biological models constructed using automata networks, as well as to evaluate their usefulness for applications.

12–1 On the Proper Use of the Networks

12–1–1 Scope of the Models

As we implied in the introduction, "large" networks (large in the number of automata) are used from the point of view of disordered systems. The models of the nervous system or the genome which we mention in this book are extremely simplistic, not only compared to biological reality, but even compared to what is currently known about these systems. Our justification is clear:

- We do not know, and we have little hope of ever knowing the complete details of the interactions between the different elements of the system.
- Even if we knew them, it would take exponentially long times to describe the space of configurations accessible to the system.

The question is then to assess the scope of models based on such coarse approximations. The justification which we offer is a conjecture based on the concept of universality classes. This idea comes from the renormalization group method applied to phase transitions. Rest assured, we have no intention of describing this method here, but only of discussing its results. Physics offers us numerous examples of phase transitions: fusion, liquefaction, transitions in magnetic systems, insulator-metal transitions, superconductivity, etc. At a particular temperature, the transition temperature, several physical quantities characteristic of the system vary in a discontinuous way. The paradox lies in the fact that the scaling laws describing the variations of these quantities near the critical point are the same for transitions of apparently very different natures, while the transition temperatures themselves appear to be unpredictable. The set of transitions which obey the same scaling laws constitutes a universality class. The renormalization group method has been used to predict these universality classes, as well as the observed scaling laws. These classes only depend on the dimensions of the space and the order parameter. One of the consequences of this universality is that all mathematical models which abide by these dimensions can be used to predict the qualitative behavior of the real system as well as the scaling laws it obeys. These notions have been generalized to include disordered systems (percolation transitions, for example) and continuous dynamical systems (transitions toward chaos).

Our conjecture, in applying these automata networks to living systems, is that the simplified mathematical description and the real system belong to the same universality class. Consequently, certain properties predicted by the model are independent of the details of the complete description of the system. Among these "universal" properties can be found the qualitative properties, such as the existence of attractors or of chaotic regimes, as well as semi-quantitative properties such as scaling laws. For a scaling law of the form:

$$m = AN^{\alpha}$$

there is a good chance that the exponent α will be correctly predicted by the theory. On the other hand, the factor A is not universal, and depends on the modeling details.

Now that we have determined what we can legitimately expect from these models, it is easy to specify the type of information that they can never give us. Given that the predicted properties are not very sensitive to the nature of the interactions, no conclusions can be drawn about the latter. The fact that a mathematical model based on certain hypotheses on the interactions among cells "works," meaning that it predicts qualitative behaviors or scaling laws confirmed experimentally, does not validate these hypotheses.

In this sense, we can consider modeling by automata networks to be complementary to the molecular biology approach, since the aim of the latter approach is precisely to uncover the exact interaction mechanisms at the molecular level.

12–1–2 Conditions of Applicability of the Algorithms

The automata network approach enables us to tackle such varied applications as signal processing, combinatorial optimization, or expert systems. For each particular case, the network used may be different, but many of the concepts are common to all of these methods. The primary advantages of this approach, other than this conceptual unity, is the relative ease of programming and the potential use of parallel algorithms.

This does not imply that this is a universal panacea, which can be applied indiscriminantly. The following warnings are imperative:

- There exists no universal algorithm which is optimum regardless of the application.
- For a particular problem, a classical algorithm can be more efficient than network algorithms. For example, the Lin-Kernighan algorithm for the traveling salesman problem still seems to be the best one.
- "Canonical" algorithms, such as the Hopfield algorithm or simulated annealing, are rarely well adapted to concrete problems of real sizes. In these cases, we tend to use better adapted variants instead.
- Recall that the implementation of an algorithm requires the choice of an architecture, an efficient coding of the data, which can require appropriate preprocessing, and the adjustment of the control parameters of the algorithm (such as the temperature for simulated annealing, or the multiplicative factors for the gradient descent method).

In other words, applying network algorithms to a given problem can give excellent results, but expertise in automata networks without a corresponding understanding of the field of application is insufficient to guarantee the success of the project.

12–2 The State of the Art and Future Perspectives

The examples given in this book are but a very incomplete sample of a large number of studies and results on these networks. A particular implementation corresponds to a set of binary choices among possibilities which we have already discussed:

digital or analog units
deterministic or probabilistic units
boolean functions or threshold functions
random or cellular connections
symmetrical or unsymmetrical connections
homogeneous or layered network
iteration in series or in parallel
a single iteration through the network or iterating until the attractor is reached

To be sure, we have not attempted to give an example of each of the possible choices, but the published literature contains a large number of them, such as:

- the use of boolean networks for recognition and memorization tasks (I. Aleksander in *Neural Computers*, 189–197, see ref. in bibliography);
- the existence of several types of layered networks (K. Fukushima and S. Mikaye, *Biol. Cyber.*, vol. 28, 201–208 (1978), B. Huberman and T. Hogg, *Phys. Rev. Let.* vol. 52, 1024 (1984));
- a model by J. Hopfield and D. Tank (in *Disordered Systems and Biological Organization* (1986), 155–170, see ref.) using analog units and a Hebbian learning rule;
- a model by C. von der Malsburg and E. Bienenstock (in *Disordered Systems and Biological Organization*, 247–272 (1986), see ref.) based on a dynamics of variable connections during the recognition process, and which has invariant recognition properties.

As for biology, modeling the immune system leads to some interesting perspectives. Here are three references:

- M. Kauffman, J. Urbain, and R. Thomas, *J. Theor. Biol.*, vol. 114, 527–561 (1985).
- G. Weisbuch and H. Atlan, *J. Phys. A*, vol. 21, L189–192 (1987).
- G. Parisi in *Chaos and Complexity*, edited by R. Livi, S. Ruffo, S. Ciliberto, and M. Bulatti (World Scientific, 1988).

In this book we have only discussed models which use "large networks." "Small" networks are also used, in cell biology, for example. See the collective book edited by R. Thomas "Kinetic Logic," *Lecture Notes in Biomathematics*, volume 29 (Springer Verlag, 1979).

For technological applications, refer to the bibliography at the end of the chapter. Let's first say a few words about problems which are of interest to computer scientists, and which we have not discussed thus far.

12–2–1 Computer Science Problems

SYSTOLIC ARRAYS Systolic arrays are networks of cellular automata which all perform the same simple arithmetic operation, enabling repetitive tasks to be carried out in parallel. For example, in the case of matrix multiplication, each unit computes the product of two of its inputs and adds the result to the third input to determine the output signal. These networks make it easy to implement dedicated (single function) processors in silicon, which are very fast because of the parallelization, and can then be invoked by a host computer. Chapter 8 of the classic book by C. Mead and L. Conway *Introduction to VLSI Systems* (Addison-Wesley, 1980), is a good introduction. The paper by Y. Robert "Systolic Algorithms and Architectures," in *Automata Networks in Computer Science* (see ref.), is more recent (1987).

COOPERATION BETWEEN PROCESSORS One of the first applications of automata networks has been to the problem of sharing resources in a computer system between the different processors and the memory resources, the printers, and the assorted peripheral devices. This problem comes up today in the design of general parallel systems MIMD (Multiple Instruction Multiple Data). Petri networks are one of the formalisms used to model these systems (G.W. Brams, *Petri Nets: Theory and Practice* (Masson, 1983).)

12–2–2 Specialized Machines

All of these parallel algorithms call for the construction of parallel machines in order to fully exploit their potential. These machines have been very slow to come on the market. The first obstacle is that the most efficient parallel machine to implement a particular algorithm would be very specific to that algorithm. However, all of these algorithms are still under development, so it hardly seems reasonable to fix the architecture of a machine today without being sure of the algorithm. Most of the current studies on automata networks are being done on classical machines. Even when we can use a vectorial machine such as the CRAY, it is always in the context of a traditional programming language, such as FORTRAN, for example. Let's nonetheless mention a few machines particularly well suited for network calculations.

PIPELINE ARCHITECTURE Pipeline architectures are very well suited for cellular automata. The pipeline is composed of a series of simple specialized processors in which information circulates from one processor to another. In short, this is a one-dimensional automata network. In the CAM and RAP machines (at MIT and at the Ecole Normale Supérieure, respectively), the states of the automata are stored in memory planes and entered sequentially in a processor which functions as an address decoder. The address calculated is that of the truth table of the cellular automaton in question. The output of the second processor is therefore the contents of this address. This signal is sent to the memory planes to update the state of the automaton and to an image processor to simultaneously visualize the state of the network.

Programming such a machine consists essentially of loading the truth tables for the different cellular automata from the host computer, which is generally a microcomputer. These machines allow networks of 256 by 512 automata to be updated and observed at television video rates (50 images per second). These machines are described in the book by T. Toffoli and N. Margolus, *Cellular Automata machines* (MIT Press, 1987), or *Physica D*, vol. 10, 185–204 (1984), and in A. Clouqueur and D. d'Humières, *Complex Systems*, vol. 1, 585–597 (1987).

MULTIPROCESSOR SYSTEMS Machines composed of a large number of identical processors connected in a network have recently been built and used to implement automata networks. The "Connection Machine" proposed by D. Hillis contains 65,536 processors connected in a 16-dimensional hypercube. It is described in *The Connection Machine*, by D. Hillis (MIT Press, 1985) or *Physica D*, vol. 10, 213–228 (1984). Along the same lines, arithmetic processors specifically designed to be used in networks, called transputers, are being produced by the INMOS company. They are described for example in "The Transputer" by D. May and R. Shepherd, in *Neural Computers*, see refs.

CHIPS AND OPTICAL MACHINES Finally, the hardiest souls undertake to build semiconductor or optical devices—in order to directly implement Hopfield networks, for example. Several experimental chips have already been built at Bell Laboratories and at Caltech (see the papers by the Bell labs group, 182–187 and 227–234, and by the Caltech group, 408–413, in *Neural Networks for Computing*, see below, or Carver Mead's recent book, *Analog VLSI and Neural Systems* (Addison-Wesley, 1989)). One of the most promising approaches is the optical computer. In fact, elementary operations such as the transmission of light signals, multiplication by synaptic weights, and summation, can be done very easily in optics by transmission through air, the passage through an absorbant medium, such as a photograph or a hologram, and focalization, respectively. The matrix of synaptic connections is implemented by a hologram where the contributions of the different patterns are superimposed. Changing a network simply amounts to changing the hologram. Nonlinear operations, such as thresholding, are carried out by matrices of nonlinear optical devices. Successive iterations only require a closed optical path achieved by mirrors and beam splitters. The great advantage of the optical computer lies in its ability to do parallel processing, which is much greater than that of a chip; and the fact that the images to be processed can be sent directly to the computer, without any intermediate optoelectronic sensors or digitization. The second part of the article by D. Psaltis, C. Park, and J. Hong in *Neural Networks*, vol. 1, 149–163 (1988) contains an introduction to these techniques. The one in the March, 1987 issue of *Scientific American*, by Y. Abu-Mostafa and D. Psaltis, is more descriptive.

12–3 Bibliography

Books

To our knowledge, there exists no complete treatise on automata networks. The following books cover various aspects of the subject.

Two "historical" books follow:

The work of von Neumann was published after his death in *J. von Neumann: Theory of Self-Reproducing Automata*, edited by A. Burks (Univ. of Illinois Press, 1966).

The classical book by M. Minsky and S. Papert on the perceptron has just been reedited:

M. Minsky and S. Papert, *Perceptrons* (MIT Press, 1988).

The following books are listed in the order of the chapters of this book.

S. Wolfram's book, *Theory and Applications of Cellular Automata* (1986), is in fact a collection of articles, like other books published by World Scientific.

T. Kohonen's book, *Self-Organization and Associative Memory*, Springer Series in Information Sciences, vol. 8 (Springer Verlag, 1988), lists a large number of applications of algebraic methods.

The collective book *Automata Networks in Computer Science*, edited by F. Fogelman-Soulié, Y. Robert, and M. Tchuente (Manchester University Press, 1987), describes many aspects which are not discussed here, from mathematical tools to systolic arrays. This book, written in a formal mathematical style, also discusses linear formalisms and back-propagation, with comparative evaluations of the different algorithms.

The collective book *Parallel Distributed Processing*, edited by J. Rumelhart, J. McClelland, and The PDP Research Group, in 2 volumes (MIT Press, 1986), is representative of the interest researchers in cognitive sciences have shown for "connectionism" (modeling cognitive systems by automata networks). Several approaches are described. The third volume (1988) contains directions for programming the different algorithms, as well as a diskette with some programs.

Another commented collection of articles, *Spin Glass Theory and Beyond*, by M. Mézard, G. Parisi, and M. Virasoro (World Scientific, 1988), is representative of physicists' approach to automata networks and combinatorial optimization. The most classical articles are included. The book by D. Amit, *Modeling the Brain Function* (Cambridge University Press, 1989), is a clear introduction to these aspects.

Finally, a book which is devoted to algorithms is *Computer Simulation and Computer Algebra*, by D. Stauffer, F. Hehl, V. Winkelmann, and J. Zabolitzky (Springer Verlag, 1988).

Conference Proceedings

The published proceedings of workshops (small conferences by invitation only) are a good way of staying informed about the subject.

D. Farmer, T. Toffoli, and S. Wolfram, eds., *Dynamical Systems and Cellular Automata* (Academic Press, 1985).

E. Bienenstock, F. Fogelman-Soulié, and G. Weisbuch, eds., *Disordered Systems and Biological Organization*, NATO ASI Series in Systems and Computer Science, vol. F20 (Springer Verlag, 1986).

J. Denker, ed., "Neural Networks for Computing," *Conf. Proceedings 151*, Snowbird, Utah, 1986 (American Institute of Physics, 1986).

R. Eckmiller and C. von der Malsburg, eds., *Neural Computers*, NATO ASI Series in Systems and Computer Science, vol. F41 (Springer Verlag, 1988).

And many more...

Journals

One of the most difficult tasks in establishing a bibliography on this subject is that the publications are scattered in very diverse journals. A few recent journals are specialized in automata networks:

Complex Systems, published by Complex Systems Publications, is oriented toward cellular automata.

Neural Networks, published by Pergamon, *International Journal of Neural Systems*, published by World Scientific, *Networks, Computation in Neural Systems*, published by the Institute of Physics, and *Neural Computation*, published by MIT Press, are oriented toward neural networks.

However, more frequently than not, physicists publish in physics journals and computer scientists in the *Proceedings of the I.E.E.E.*

APPENDIX

Algorithms

The aim of these notes is not to cover in detail all of the algorithms used to iterate automata networks and study their dynamics, but rather to give the necessary background needed to begin to program on a microcomputer the examples discussed in the first four chapters. Numerical simulation is the favorite approach in automata networks; it plays somewhat the same role as experiments in other fields. The phenomena are first observed with the help of simulations, and are then interpreted, if possible, by the theory.

Coding

Configurations

Generally speaking, the decimal code $d(c)$ of a configuration c of k automata denoted by the subscript j is written:

$$d(c) = \sum_{j=0}^{k-1} 2^j e(j)$$

where $e(j)$ is the state of the automaton subscripted by j. Conversely, the state configuration of automata $e(j)$ is the binary code for d. The correspondence is bijective. The $e(j)$ are the remainders of the successive divisions of d by 2. Table 2.2 gives the decimal codes and the corresponding binary configurations from 0 to 31.

Boolean Functions

We start by coding the input configurations using the decimal code. In the general case, if the decimal code of each input configuration, denoted d, varies from 0 for the $000\ldots0$ configuration to 2^k-1 for the $111\ldots1$ configuration, the code f of the boolean function is given by:

$$f=\sum_{d=0}^{m-1} 2^d o(d)$$

where $o(d)$ is the output which corresponds to the input configuration d, and $m=2^k$.

As an example, we code the 16 boolean automata with two inputs from 0 to 15. We choose as the most significant bit the one which corresponds to the input configuration 11 and as the least significant bit the one which corresponds to 00. The AND function, for example, has a code of 8:

- code(AND) $= 0\times2^0+0\times2^1+0\times2^2+1\times2^3=0\times1+0\times2+0\times4+1\times8=8$

and the OR function, 14:

- code(OR) $=0\times1+1\times2+1\times4+1\times8=14$.

The exclusive OR function has a code of 6, and the NAND, 7. Table 2.1 gives the decimal codes and the truth tables of the 16 boolean functions with two inputs. Some examples of the coding of boolean functions with three inputs can be found in chapter 3, section 3.2. These codes allow the functions of the boolean automata to be written in a simple and standard way, as well as to be easily stored and manipulated in the computer.

Definition of the Network

A network is defined by:

1. Knowing the *inputs* of each automaton.

In the general case, for an arbitrary connectivity, each input of each automaton is given by a matrix: in the case of automata with two inputs, two matrices, called for example $e1$ and $e2$, define input automaton #1 and input automaton #2 for each of the automata.

For the example of figure 2.4:

$$e1(1)=2,\ e1(2)=3,\ e1(3)=4,\ e1(4)=5,\ e1(5)=1;\ \text{and}$$
$$e2(1)=3,\ e2(2)=4,\ e2(3)=5,\ e2(4)=1,\ e2(5)=2.$$

In general, it is not necessary to define the outputs of each automaton.

In the case of cellular connectivity, it can be simpler to recalculate the input automata each time rather than to store them in memory.

2. The choice of the *transition rules* of the automata.

In the case of boolean automata, the result of the transition rule must be given for each automaton and for each input configuration. We then use a matrix with subscripts i and j, where one subscript refers to the automaton and the other to the input configuration. This second subscript is the decimal code of the input configuration, incremented by 1. For two inputs:

$$j = 1 + \text{eta}(e1(i)) + 2 \times \text{eta}(e2(i))$$

where eta gives the state of the automaton. We add a 1 so that j varies from 1 to 2^k, where k is the connectivity of the network.

3. The choice of the *mode of iteration* (see the corresponding programming principle below).

The Initial Conditions

1. Initially, the time is set to 0.
2. The *initial configuration* of the network must be chosen. Depending on the application, it can be:

- programmed, for simple cases of a homogeneous configuration, or of a random configuration (in the case of an exhaustive study to determine the iteration graph, the decimal code of the configuration is the index of the initial condition loop),
- input from the keyboard, using a text editor or a graphical program, or
- finally, it can be determined by external data gathered by a set of sensors, in the case of an application to image or pattern recognition.

Iterations

The central part of the program is the iteration of the transition functions. This is what takes the most execution time, even though the number of programming lines is often quite small. Consequently, this is the part of the program that is most worthwhile to optimize. An external loop increments the time. An internal loop, subscripted by i, the number of each automaton, gives the new state of the automaton as a function of the states of the connected automata at the preceding time.

After having evaluated j, the code of the input configuration of automaton i according to the above equation, its new state, neta(i), is calculated by the transition function $f(i,j)$:

$$\text{neta}(i) = f(i,j)$$

There are two cases:

- For iteration in series, the state of the automaton is updated directly and we proceed to evaluate the following automaton. The preceding equation can therefore be written:

$$\text{eta}(i) = f(i,j)\,.$$

- For a parallel iteration, the new states of all of the automata must have been calculated before the input configurations can be modified. Two iterations on the automata must then be performed, the first to calculate all of the new states (matrix neta(i)), and the second to update the matrix of the states which will be used to calculate the input configurations for the next iteration:

$$\text{eta}(i) = \text{neta}(i)\,.$$

Determining the Iteration Graph

The idea behind this method is to *color* the nodes of the iteration graph. We use a "graph" vector, whose subscript is the decimal transcription of each of the 2^N configurations. (The size of this matrix is the limiting factor for this method.) To each element of this matrix will be assigned a whole number specific to the basin of attraction, the *color* of the node.

We initially set the color of the matrix to 0. Starting from the first configuration, we iterate the state change functions, placing 1's in the elements of the vector graph which correspond to the nodes which have been reached. Of course, we look at the state of each node before changing it: if it is equal to 1, the limit cycle has been reached.

The algorithm continues, incrementing the color of the nodes each time we look for a new basin, starting from the smallest configuration whose color is still 0. The iteration of the state change functions stops each time the color differs from 0: we have found a new basin of attraction if the color found is the same as that of the initial condition we started from. If the color is smaller, we are falling into a basin that has already been found. The preceding color must therefore be reassigned, starting from the last initial condition. The retrace operations are faster if we keep a successor matrix which stores the successor of each configuration. Similarly, during the retrace, one can increment a counter at each step to measure the periods.

The algorithm is finished when all of the nodes are colored. The highest color gives the number of attractors.

The Statistical Approach

If the number of automata is greater than about thirty, the preceding algorithm, which requires a memory size of the order of 2^N and a time of the order of $N2^N$, is inapplicable. The statistical method consists of starting from randomly chosen initial conditions and collecting dynamical data.

To find a limit cycle, a reference must be stored in memory: this is the configuration reached by the network after an arbitrary time which is assumed to be greater than the transient period. The algorithm consists of taking the reference at time t_{ref}, and comparing it to all of the configurations obtained. A counter can be used to measure the period. It is possible that the reference may have been taken too soon: in fact the comparison is done between times t_{ref} and $2 \times t_{ref}$. If the limit cycle has not been detected at $2 \times t_{ref}$, we take a new reference at $2 \times t_{ref}$ and we multiply the search time by 2. The operation "new reference–new search" continues until the limit cycle is reached, unless we decided *a priori* to interrupt the search at a maximum time.

To estimate the number of attractors, we must compare the configurations reached during a limit cycle to the configurations obtained for a previously determined limit cycle. If number of attractors is high, this can be a long search. This is why certain decimal codes are sometimes used.

Cellular Automata and Boundary Problems

The algorithms proposed in the general case can of course be applied to cellular automata. They can be simplified by taking advantage of the regularity of the connections and the uniformity of the state change functions.

It is not necessary to keep connection matrices representing the inputs of the automata, since each matrix can generally be obtained by a simple translation from the position of the automaton under consideration. A small difficulty comes up due to the finite character of the lattice used in simulations. Cellular automata are defined for infinite lattices, whereas the simulated lattices are limited by boundaries. There are several ways of dealing with boundary effects. One way, for example, is to fix the states of the automata on the boundaries to arbitrary values. The most frequently used solution consists of connecting opposite edges to each other, which has the advantage of being less rigid than the previous solution. A linear lattice then forms a circle, and a higher-dimensional lattice forms a torus or a hypertorus.

To take into account the conditions at the edges, we add two "surfaces" to the lattice (in fact, an automaton at each edge for one dimension, two rows of automata in two dimensions, etc.). The iteration loops which calculate the new state, neta(i), only apply to the internal lattice. The states of the edges are only updated during the second loop, which updates the vector eta(i); and they then take on the values of the new eta calculated for the "real" edge to which they are connected. For example, if a linear lattice of 10 automata is being simulated, these automata are numbered from 2 to 11. We add to them a left edge, numbered 1, and a right edge, numbered 12. At the beginning of any loop from 2 to 11 to update neta(i), eta(1) is equal to eta(11) and eta(12) is equal to eta(2). In this way, the state of automaton 11, the farthest to the right, is set as a function of a right-hand signal sent by automaton 2, which is the farthest to the left. Of course, if the connections involve more than the nearest neighbors, the size of the edges is correspondingly increased.

Regardless of the dimension of the lattice, each automaton can be referenced by a single integer. For a cube with dimensions L, M, and N, the subscript i of the automaton in a vector eta, for example, is then given by:

$$i = u + v \times L + w \times L \times M \, ,$$

where u, v, and w are the coordinates of the automaton in the cube. It follows that i goes from 1 to $L \times M \times N$. Its six nearest neighbors have subscripts: $i \pm 1$, $i \pm L$, $i \pm L \times M$. If this representation is chosen, it is then convenient to update the edges by interconnecting the edges with a single loop. For the preceding cubic lattice, $L \times M$ automata are placed in the first position, which constitutes the left side. Then, the automata whose subscripts go from $L \times M + 1$ to $L \times M \times (N+1)$ constitute the real lattice. Finally, $L \times M$ automata are added to the right edge. It follows that i goes from 1 to $L \times M \times (N+2)$. neta(i) is iterated for i going from $L \times M + 1$ to $L \times M \times (N+1)$. The left edge is then updated according to:

$$\text{eta}(i) = \text{eta}(i + L \times M \times N)$$

for i going from 1 to $L \times M$. The equivalent procedure is used for the right edge. In fact, this method is equivalent to deforming the cube into a super-helix rather than into a torus. It has the advantage of being programmable in a single loop, regardless of the dimension of the lattice, and the size of the approximation compared to an infinite lattice is the same as for the cube.

If the lattice is one-dimensional, its successive different configurations can be displayed one after the other, according to Wolfram's representation. Rather than "generating" boring tables of 0's and 1's, it is more readable to code 0's by blanks or dots and 1's by pluses + or asterisks *. Word processing printers can sometimes cause problems if they print proportional characters: a constant spacing must be set so as to not scramble the output table. Of course, we can also choose a graphical mode of representation, which has sometimes been done in this book, representing the states of the automata by white or black dots.

If the lattice is two-dimensional, a powerful computer is needed to observe the output in "real time"; otherwise visually we can only see a sweep across the screen corresponding to the sequential update of the pixels. By using certain codes based on an alphanumeric representation, or on the color, we can integrate information about several time intervals within a single table (see figures 4.5 or 10.4).

The Gradient Descent Method

This is the classical method which is used to find the minima of a function of several variables.

Assume we want to find the minimum of y, a function of a single variable:

$$y = f(x)\,.$$

The idea is to start from an arbitrary point x_0 (or a cleverly chosen point when it is possible) and to move in successive steps toward the minimum. The calculation of the displacement from a position x_n to the following position x_{n+1} depends on the slope of the function in the neighborhood of x_n. The steeper the slope, the larger the displacement; a shallow slope indicates that we are near the minimum, whereas a steep slope indicates that there is still a long way to go. The algorithm consists of moving at each step according to:

$$x_{n+1} = x_n - \varepsilon f'(x_n)$$

where ε is a carefully chosen multiplicative coefficient. At each step we test the derivative $f'(x)$ of $f(x)$ and we stop the iteration when $f'(x)$ is sufficiently close to 0. In fact, one of the difficulties of this method rests in the choice of ε. If ε is too small, a large number of iterations are needed before a reasonable approximation to the minimum is obtained. If ε is too big, we are in danger of oscillating around the minimum or of finding ourselves in a different valley, and thus losing track of the minimum we were looking for.

In the case of a function of several variables, x, y, z,..., the method can be generalized by using the gradient of $f(x, y, z, \ldots)$. The iteration formula then becomes:

$$\begin{aligned} x_{n+1} &= x_n - \varepsilon f_x{}'(x_n, y_n, z_n, \ldots) \\ y_{n+1} &= y_n - \varepsilon f_y{}'(x_n, y_n, z_n, \ldots) \\ z_{n+1} &= z_n - \varepsilon f_z{}'(x_n, y_n, z_n, \ldots) \end{aligned}$$

The notation $f_x{}'$ represents the partial derivative of $f(x, y, z, \ldots)$ with respect to x. The displacements in the space of the variables are then proportional to the gradient of the function f and take place along the line of steepest slope. Of course, the difficulties concerning ε are the same as for the one-dimensional case.

Index

G

H

I

L

M

N

O

P

R

S

T

U

W